C0-AWR-529

POETIC LICENSE

POETIC LICENSE

Authority and Authorship in Medieval and Renaissance Contexts

JACQUELINE T. MILLER

New York Oxford
OXFORD UNIVERSITY PRESS
1986

Oxford University Press
Oxford New York Toronto
Delhi Bombay Calcutta Madras Karachi
Petaling Jaya Singapore Hong Kong Tokyo
Nairobi Dar es Salaam Cape Town
Melbourne Auckland

and associated companies in
Beirut Berlin Ibadan Nicosia

Published by Oxford University Press, Inc.,
200 Madison Avenue, New York, New York 10016

Library of Congress Cataloging-in-Publication Data

Miller, Jacqueline T.
Poetic license.

Bibliography: p.
Includes index.
1. English poetry—Early modern, 1500–1700—History and criticism. 2. Authority in literature. 3. Authorship in literature. 4. Chaucer, Geoffrey, d. 1400. Hous of fame. 5. Spenser, Edmund, 1552?–1599. Faerie queene. 6. Sonnets, English—History and criticism. I. Title.
PR535.A88M55 1986 821'.009 86-8686
ISBN 0-19-504103-8 (alk. paper)

2 4 6 8 9 7 5 3 1

Printed in the United States of America
on acid-free paper

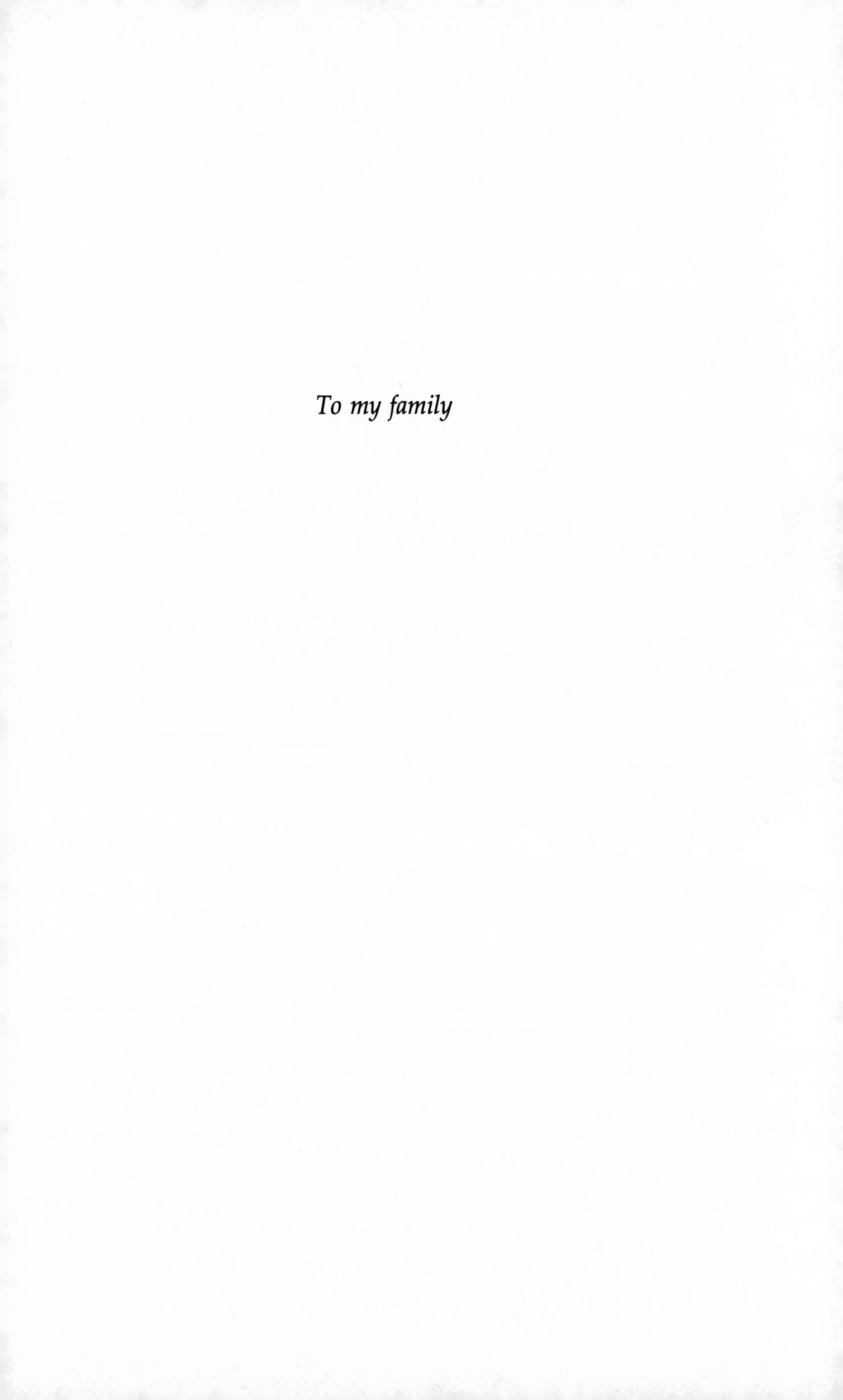

To my family

Acknowledgments

This book was completed with the aid of a fellowship from the American Council of Learned Societies. Earlier versions of portions of the book appeared as articles in *ELH* (Vol. 46, 1979), *Chaucer Review* (Vol. 17, 1982, Penn State Press), and *Spenser Studies* (Vol. 5, 1985, AMS Press), and I wish to thank the editors and publishers for permission to reprint.

Stanley Fish and Arnold Stein saw this book through its first writing at an early stage, and I am grateful for their guidance then and for their continued help and encouragement. Among the many friends and colleagues who provided support and advice, I would like in particular to express my thanks to Robert Wind and William Cain.

I save the last words of these acknowledgments for my husband, Thomas Cartelli.

Contents

POETIC LICENSE

Introduction

Authority and authorship are sometimes complementary, sometimes conflicting concepts, and the motives and strategies that work to merge or separate them take various, complex forms. Their complexity mirrors the difficult relation that exists between a writer's desire for, on the one hand, individual authority or creative autonomy and, on the other hand, the authoritative sanction that external sources provide. Authority, both when it resides with the author and when it does not, implies restraint as well as freedom, limitation as well as power.[1] A claim of personal authority may liberate and validate an author's activities; it may also restrict them, since it carries with it a constraining burden of responsibilities and is often acquired through an act of submission. Conversely, an external authorizing principle may threaten the writer's position, leaving him little or no space in which to function; yet his representation of something different from himself may be what motivates and enables him to write. My phrase "resides with the author" simplifies an intricate notion. The author may accept authority that is conferred upon him; he may simply posit and assert his authority; he may deny or abdicate his authority; he may create an audience that in turn agrees to bestow authority upon him; he may create an audience that claims authority as its own and therefore challenges his. The concept often takes more subtle forms: denials of artistic competence may defy external standards of composition; challenges to authorship, deliberately invoked, may meet with self-confident gestures of surrender; the demands of creative autonomy often simultaneously embrace and resist authoritative models. Furthermore, a poet may disclaim his autonomy only to provide himself with a beneficial type of authorial anonymity: the formal acknowledgment of a higher authority may allow the imagination to roam freely under the guise of authoritative sanction.

Authority is not, therefore, always available or even desirable for the author. It may be perceived as an inevitable concomitant of the poet's position, which he may eagerly embrace or from which he may futilely try to escape. Alternatively, it may be presented as necessary but unattainable, always sought but never achieved or found. The poet's relation to authority usually seems complicated by a pull in two directions: does he rest upon his status as a reflector of an already authorized truth, a spokesman for something other than himself, or do his status and function reside in the exercise of his own creative power? The poet who accepts or insists upon creative independence and either proceeds upon this assumption or works to validate it may increasingly discover constraints that inhibit or belie his assertion of authority. Yet the author who submits to an authority other than himself often finds it insufficient to accommodate his own vision and voice and judges the consequently necessary self-limitation as too heavy a cost.

Although I have phrased these issues, particularly in the last paragraph, in terms of dichotomies (authority or none; creative autonomy versus fidelity to what is already authorized), writers of the Renaissance and the Middle Ages rarely express the problems as clear oppositions in either their poetry or their theoretical writings. Instead, the author often attempts to maintain a delicate balance between various untenable alternatives, and the resulting tensions imbue and often generate the work. This study explores the techniques by which an author may test those alternatives or try to reach that balance and examines the informing and formative influence of those tensions.

Throughout this study my concern is with *literary* authority,[2] and I use the term *authority* to refer to that which can sanction and certify the poetic text. In this context, an authority external to the poet encompasses a wide range of possibilities: it may refer to a traditionally accepted system of belief, a fixed principle of order, a recognized figure with authoritative status (God, or a god [Jove], or a goddess [Fame or Nature]), literary conventions or traditions (a genre, or a conventional framework), literary predecessors (Chaucer's "olde bokes" and *auctores,* or the writers on whom Renaissance poets based their theory of imitation), or the structure of the actual or the natural. In short, it includes various established principles, systems, or sources that the poet cannot claim to have produced himself and that may be called upon to sanction a text that has conformed to them. I have used the term in such a wide-ranging sense because the poets may appeal to any or all of these authorities when the power of their own voices falters; moreover, it is when these sources or guides fail to function as adequate models that the individual voice emerges to be tested as its own principle of authority. These are the established criteria against which the poet's voice must certify itself when it wants to demonstrate its autonomy. When I speak of the poet's own authority, I mean his ability to create and endorse independently his own vision in his poetry, to uphold his full responsibility and power to validate that vision.

I have chosen to examine these issues in several medieval and Renaissance contexts because these periods provide significant early and fertile—but often neglected—ground for the study of the relationship between authority and authors. Studies like Harold Bloom's *Anxiety of Influence* and W. J. Bate's *Burden of the Past and the English Poet* have considered the issue primarily as one of writers' coming to terms with their literary predecessors; my broader definition of authority, while it includes this element, extends the scope of such works. But even more significant is that, within the bounds of their more limited identification of the kind of authority with which poets contend, these studies associate the problem primarily with post-Renaissance writers. Bate investigates the burden of the past on authors from 1660 to 1830; this period, he claims, provides the "first major example of this dilemma."[3] Bloom reveals a similarly modern bias by contending that the anxiety of influence is an issue for post-Miltonic writers alone.[4] Even Edward Said, whose idea of authorial authority and its "molestation" is closer to mine, focuses on writers from the eighteenth through the mid–twentieth century.[5] This insistence that the problem is a privilege of Enlightenment and post-Enlightenment writers leads not only to a facile dismissal of medieval and Renaissance writers from discussion but also to a reductive reading of the struggles that inform their works.[6] This tendency to exclude the earlier periods from the "modern" dilemma provides a distorted view, which works like Alice Miskimin's *Renaissance Chaucer* and, more recently, John Guillory's *Poetic Authority* and David Quint's *Origin and Originality in Renaissance Literature* have begun to amend.[7] One purpose of this study is to help return the Middle Ages and the Renaissance to the mainstream, to show that the tension between the desire for creative autonomy and the pressure of inherited or conventionally accepted authoritative systems or voices is a central artistic concern whose roots extend further back than is customarily acknowledged.

To this end, my concern has been not so much to isolate the distinguishing features of the two different periods under consideration as to identify and establish the general structure of the problem from a variety of perspectives and through a variety of techniques for dealing with it. My method has been to consider particular medieval and Renaissance authors as well as particular genres and modes popular in those periods: in other words, to focus not only on the issues as they are shaped by individual writers but also on the form they take in various poetic frameworks. My opening chapter discusses some of the historical and theoretical dimensions of the topic, suggesting several of the principles that may determine and deter a writer's privilege to authorize his own text. I then examine three poetic contexts, devoting a chapter to each. The first context is one in which the poet continually searches for, but fails to find, a reliable authoritative model, and thus is continually motivated to test the (equally unreliable) authority of his own voice. To set up these problems and define the terms, I analyze the medieval dream vision (focusing on Chaucer's *House of Fame*) as a form generated by an authorial stance that alternately seeks to hide behind, and asserts its superiority to, traditionally accepted sources of order and truth. In the second context, the poet initially (though not finally) assumes his power to create and sanction his own poetic vision in his poetry, presenting his text as an autonomous created fiction with an acknowledged absence of external authorization. In this chapter I discuss the problem of authority in allegory, and I explore in depth Spenser's attitude toward external models and his own form of expression in *The Faerie Queene*. The final context concerns the exploitation of these issues. This last chapter focuses on the sonnets of Sidney and Spenser and their relation to Herbert's lyrics. Here I investigate the Renaissance concept of imitation and the way problems of creative autonomy and authoritative sanction can be employed as rhetorical techniques that mediate between the speaker and the object of his poetry.

Although these various (though certainly not exhaustive) contexts reveal a range of different motives and strategies for dealing with problems of authority and authorship, they also reveal some similarities—similarities that cut across the boundaries of different literary periods, different literary modes, and different authors. There are similar gestures of denial, sometimes serious and sometimes disingenuous: Chaucer's "the book of seyth," Spenser's "as I have found it registred of old," and Herbert's "copie out onely that." There are also similar moments of assertion, from Chaucer's statement that "none other auctour alegge I" to Sidney's proposition that the poet has "all . . . under the authoritie of his penne." Yet whatever the attitudes and techniques brought into play, authority—located either within the author or outside of him—seems to be, at best, a temporary quality that lends an equally temporary legitimacy to any poetic endeavor. There is no reliable, absolute source of sanction that a poet may call upon, and the competing pull of various alternatives that surface when the others fail often provides the tension that structures the poem and the energy that produces it.

CHAPTER I

Authority and Authorship: Poetic License

I

The twelfth-century humanist and educator Bernard of Chartres reportedly claimed that "we are as dwarfs perched upon the shoulders of giants" ("nos esse quasi nanos gigantium humeris insidentes").[1] This phrase became a medieval commonplace that remained popular well through the seventeenth century, appropriated from its inception by participants in the quarrel of the ancients and the moderns. This battle has roots in antiquity and has persisted, in various forms, into modern times; it has been waged in the humanities and in the sciences; and it is, essentially, in Gilbert Highet's words, a war "between originality and authority."[2] Bernard's formulation of this issue and its adaptation by his medieval contemporaries and their Renaissance successors demonstrate one way in which the tension between the desire to assert authorial independence and the pressure to acknowledge external authorizing principles was expressed in these periods.

Bernard's remark was quoted frequently by his followers, and literary critics have tended to regard the aphoristic commonplace as testimony to the medieval habit of accepting the authority of antiquity, the superiority of the *auctores*, and, consequently, the relative insignificance of contemporary authors as originators, creators, and makers in their own right. Bernard's dictum has been called "a ringing challenge . . . to twelfth-century scholarship . . . relating to the value of the writers of antiquity"; others have claimed that John of Salisbury, by reciting it in his *Metalogicon*, "dismisses the claim of the *moderni* in his time to superiority over the ancients," and that Alain de Lille, who alludes to the saying in the prose Preface of his *Anticlaudianus*, "rejects the modern manner" of those who "consider themselves superior to the 'ancients.' "[3] A recent critic seems to summarize this opinion when he writes, "No 'modern' writer could decently be called an *auctor* in a period in which men saw themselves as dwarfs standing on the shoulders of giants, i.e. the 'ancients.' "[4]

But such readings obscure the complexity of Bernard's statement. Does he so simply dismiss the claims of the moderns to superiority? Does his comment relate only to the value of the ancient writers, and not as well to the value of the moderns? For if the comparison of dwarfs to giants implies the larger stature and hence importance of the ancients, the idea of the dwarfs' being on the shoulders of the giants suggests that the moderns have advanced to a higher position.[5] As Brian Stock has noted, Bernard considered his own age to be "a continuation of the classical world in faithfully reproducing its concepts, styles, and cultural ideals," but he also "was prepared to grant that in other respects it had perhaps surpassed even the ancients."[6] Indeed, if we look at the remark as it has been preserved for us in John of Salisbury's *Metalogicon*, we see that, in elaborating his dictum, Bernard acknowledged the twofold nature of the comparison:[7]

> Bernard of Chartres used to compare us to [puny] dwarfs perched on the shoulders of giants. He pointed out that we see more and farther than our predecessors, not be-

> cause we have keener vision or greater height, but because we are lifted up and borne aloft on their gigantic stature.[8]

Moderns may not be more acute—or taller—by themselves, and they owe a great debt to the ancients who lift them aloft, but they do see more and farther from their higher and exalted perspective. This is neither a wholesale advocacy of reverence for the ancient author nor a wholesale dismissal of the contemporary author. It attempts to acknowledge the value of the ancients and the reliance of the moderns upon them, and simultaneously to recognize the value of the moderns and the contributions their greater vision may make. The remark grants the importance of studying the ancients, but not at the cost of ignoring the achievements possible for the moderns. At stake, ultimately, is the question of authority: is the work of the modern authorized solely by the ancients, or can the modern, with his new vision, claim any authority of his own?

John of Salisbury's own context for Bernard's remarks in the *Metalogicon* delineates, as it grapples with, this issue. The school of Chartres, of which John has been called "the finest flower," has long been recognized for its humanist stance and for the significance of its pedagogical practices as a center of learning.[9] John's *Metalogicon* provides an account of Chartrian pedagogy and also (though less directly and comprehensively) addresses political-ecclesiastical issues of his time.[10] It announces itself as a polemic arguing against the "Cornifician" attack on the arts that occurred during the first half of the twelfth century; Cornificius refers to an unidentified detractor of Virgil and the arts, whose followers John locates in the monasteries, in various professions, and in the court. According to John, the Cornificians disparaged the study of *auctores*, of grammar and rhetoric, in favor of an exclusive (and, to John, empty) emphasis on dialectic. Yet despite John's rejection of the Cornificians, he remains a strong proponent of the dialectical method (as is clearly expressed in the last two books of the *Metalogicon*),[11] as well as a sometimes cautious spokesman for the recognition of the moderns (as my discussion below will elucidate).[12]

In the *Metalogicon*, John introduces Bernard's comparison of dwarfs and giants with a similar statement in his own words:

> Our own generation enjoys the legacy bequeathed to it by that which preceded it. We frequently know more, not because we have moved ahead by our own natural ability, but because we are supported by the [mental] strength of others, and possess riches that we have inherited from our forefathers.

And he adds, after reporting Bernard's comment, "I readily agree with the foregoing" (III.4; p. 167). But with what does he readily agree? With two propositions—both that the moderns see better than the ancients and that they can do so because they are supported by the ancients. He begins this chapter of the *Metalogicon* by speaking of the usefulness of Aristotle's *De Interpretatione*, as well as of its difficulty.[13] Almost apologetically, he suggests the superiority (both in style and content) of the modern authors on this subject:

> . . . if I may say so, begging leave of all, any one of the doctors could (as many of them in fact do) more concisely and lucidly provide everything that is taught in this book in the elementary lessons which they call *Introductions*. The only thing lacking would be the respected authority of the [author's] words. There is hardly any [of the doctors] who would not, in addition to teaching what is contained in this book, also add other things equally necessary, without which a knowledge of the art cannot be acquired.
>
> (III.4; pp. 165–66)

And he proceeds to cite Abelard's comment that a contemporary could easily

> compose a book about this art, which would be at least the equal of any of those written [on the subject] by the ancients, in both its apprehension of the truth and the aptness of its wording, but [at the same time] it would be impossible or extremely difficult for such a book to gain acceptance as an authority.

According to John, then, the work of a modern may equal and even excel that of any particular ancient, but none-theless may be denied the authoritative status and reverence that has accrued to the older and earlier writer. However, here he agrees with Abelard that the ancients *deserve* such status and regard, because it is their combined hard work that enables the moderns to pursue their accomplishments:

> He also used to assert that recognition as authorities should be conceded to these earlier authors, whose natural talent and originality flourished in fertile luxuriance, and who bequeathed to [an indebted] posterity the fruits of their labors. . . .
>
> (III.4; p. 167)

Yet elsewhere, John also insists that the moderns be given their due status and be acknowledged and cited as authorities when appropriate. In his Prologue to the *Metalogicon*, he firmly states his position on this matter:

> I have not been ashamed to cite moderns, whose opinions in many instances, I unhesitatingly prefer over those of the ancients. I trust that posterity will honor our contemporaries, for I have profound admiration for the extraordinary talents, diligent studies, marvelous memories, fertile minds, remarkable eloquence, and linquistic proficiency of many of those of our own day.
>
> (Prologue; p. 6)

He reiterates this view in even stronger terms as he begins the third book of the *Metalogicon*, which contains the reference to Bernard. First he claims that a true statement should carry the same weight if said by a modern as if said by an ancient—and that no statement should be considered authoritative or not simply because of the age of its source:

> Something that is true in itself does not melt into thin air, simply because it is stated by a new author. Who, indeed, except someone who is foolish or perverse, would consider an opinion authoritative, merely because it was stated by Coriscus, Bryso, or Melissus? All of the latter are alike obscure, except so far as Aristotle has used their names in his examples. And who, except the same sort of person, will reject a proposition simply because it has been advanced by Gilbert, Abelard, or our own Adam?
>
> (III.Prologue; p. 144)

Next he proceeds not only to chastise those who disparage the moderns and refuse them potential status as *auctores* but also to suggest that the moderns may deserve that status more than the ancients:

> I do not agree with those who spurn the good things of their own day, and begrudge recommending their contemporaries to posterity. None of the latter [none of our contemporaries] has, so far as I know, held that there is no such thing as a contradiction. None of them has denied the existence of movement and asserted that the stadium is not traversed. None of them has maintained that the earth moves because all things are in motion, as did Heraclitus, who, as Martianus puns, is red hot, because he is all afire, since he maintains that everything was originally composed of fire. But these opinions of the ancients are admitted, simply because of their antiquity, while the far more probable and correct opinions of our contemporaries are, on the other hand, rejected merely because they have been proposed by men of our time.
>
> (III.Prologue; pp. 144–45)

John insists that it is only folly or perversity that prevents the recognition of moderns as authorities in their own right.[14]

Just as, then, John can defend grammar as well as logic and dialectic in the *Metalogicon*, so can he attest to the giantesque stature of the ancient *auctores* as well as affirm and applaud the accomplishments of the modern dwarfs who rise to greater heights on their shoulders. In a way, the combination of John's respect for the grammatical-rhetorical tradition and his support for dialectic is mirrored in his double-edged use of Bernard's comment—to honor the ancients and praise the moderns. Yet a major point of tension and ambivalence arises for him concerning the question of authority. Whereas John can approvingly quote Abelard's statement that the ancients deserve the authority attributed to them and denied to the moderns, he also frequently calls this position arbitrary and insists upon his right and responsibility to cite the moderns as equal or greater authorities, criticizing the bias against them. Indeed, in his hopes and recommendations that posterity will honor his contemporaries, he implicitly suggests that the writers of his age will become the *auctores* of the next.[15]

Succeeding generations also recognized the twofold nature of Bernard's famous phrase. Renaissance examples of it abound, and the evolution of the phrase reveals a growing emphasis on the stature of the modern. In fact, the granting of honor and authority to the modern seems to be the primary impulse behind references to the twelfth-century aphorism in the Renaissance.[16] As we have seen, however, this trend is neither a radically new departure from the older usage nor an exploitative distortion of it.[17] Rather, it develops directly out of the dual meaning that was attached to the comparison almost from its inception. Nor is this later rendering free of ambivalence and tensions similar to those that characterized the earlier usage. Robert Burton, who in his *Anatomy of Melancholy* provides one of the most oft-quoted examples, is a good case in point. In his Preface, Burton has Democritus Junior explain to his reader that

> Though there were many giants of old in physic and philosophy, yet I say with Didacus Stella, "A dwarf standing on the shoulders of a giant may see farther than a giant himself"; I may likely add, alter, and see farther than my predecessors.[18]

On the surface a bold and self-assured statement of superiority, Burton's words emphasize the advancement of the modern over the "giants of old." Yet he still acknowledges the giantesque stature of his predecessors, his own relative smallness, and his position on their shoulders, which enables him to see farther. Whereas Bernard of Chartres claimed that a modern dwarf can see farther than the giants of old because he is lifted up on their shoulders, Burton claims that the modern dwarf lifted up on the shoulders of the giants of old can see farther. In other words, while the syntactic emphasis of the comparison has changed since John of Salisbury's rendering of Bernard's dictum, the duality remains. Moreover, Burton's own context for this statement highlights its ambivalence. Burton has been defending—and apologizing for—his own method of extensive borrowing from other authors.[19] As though expecting an attack from readers, he justifies himself by explaining that while he has taken material from diverse writers, he has, in collecting them together, made a "new bundle of all" (p. 24); furthermore, he has cited his sources, giving "every man his own" (p. 25), and so cannot be accused of stealing. Finally, he insists that a writer has no other option. "We can say nothing but what hath been said," he remarks; only the method and style change with each succeeding author (p. 25). It is here that Burton brings in the analogy to dwarfs and giants, proclaiming his ability to see farther than his predecessors. Thus, in the paragraph introducing the allusion, Burton has acknowledged the superiority and authority of the writers he cites, claiming that he gives them due honor by naming them, and insisting that nothing new can be said; he also has announced, however, that he has made a "new bundle" out of his various sources and that the matter "becomes something different in its new setting" (p. 25). In short, Burton, when justifying his use of his predecessors, wavers between claiming that his work consists of and is authorized completely by the works of others (and that no alternative exists), and claiming that there is something "new" and distinctive in his work that is to be attributed to him as author. Furthermore, the two most conspicuously conflicting statements in this section—that "We can say nothing but what hath been said" and that "I may likely add, alter, and see farther than my predecessors"—are both duly quoted and attributed to other authors: the first, to Johann Wecker, a sixteenth-century Swiss physician, and the second, as already noted, to Didacus Stella, the sixteenth-century Spanish exegete. Burton, in other words, cites authorities not only to justify his reliance on them but also to confirm his ability to depart from them. Thus, as he acknowledges the authority of his authors, he also undermines it.[20]

This is all in keeping with an author who, calling himself Democritus Junior, chooses a persona that associates him with an ancient Greek scholar and philosopher, and then denies that he does so—as others have—to gain authority from the association.[21] He admits to "arrogating another man's name," claiming that he "would not willingly be known," and then, in final explanation of his use of "this usurped name" (p. 15), provides two possible reasons: first, that he intends "in an unknown habit to assume a little more liberty and freedom of speech" (p. 19), and second, that he intends to "imitate," "*quasi succenturiator Democriti* [as a substitute for Democritus]" the treatise he began (p. 20). He masks himself, in short, as the successor of an ancient Greek because he wants to follow the ancient's footsteps and because he wants an anonymity that will allow freedom of expression for his own voice. One reason honors the authority of the *auctores* by declaring that the ancients are worthy of being imitated, and the other quite candidly exploits it by declaring that the sanction and cover of their names allows the individual to speak freely for himself.

Burton's ruminations about his own authorship and its authority in the *Anatomy* are documented not only in the ambiguity of the narrative relationship between him and Democritus Junior but in the history of the text's editions as well.[22] In the first edition (1621), Burton provides a conclusion in which he unmasks Democritus and to which he signs his own name, acknowledging his authorship. By the second edition (1624), he has discarded this conclusion, though he interpolates much of its content into the Preface of Democritus Junior. Burton's name is absent from all subsequent revised editions. Beginning with the third edition of 1628, however, the title page, which names Democritus Junior as the author, also carries a picture of Burton himself that is identified (by caption) as a picture of Democritus Junior. We can observe in these revisions Burton's competing impulses to mask and to reveal himself as Democritus. The denial of his own authorship and the association with another authority provide him with a freedom of personal expression, yet he cannot resist asserting his authorship in subtle ways, and at one point he even denies any desire for the sanction of this other authority.[23] At the same time, however, he tries to avoid the responsibility of authorizing his own work, and toward the end of the Preface he offers as an excuse for what he writes, " 'Tis not I, but Democritus, *Democritus dixit*," reminding his readers to consider the difference between speaking "in one's own . . . person" and speaking in "an assumed habit and name" (p. 121). Finally, he associates his assumed persona with a period of history when there was complete freedom of speech:

> Object then, and cavil what thou wilt, I ward all with Democritus' buckler . . . *Democritus dixit,* Democritus will answer it. It was written by an idle fellow, at idle times, about our Saturnalian or Dionysian feasts, when, as he said *nullum libertati periculum est* [there is no danger to liberty], servants in old Rome had liberty to say and do what them list. . . . The time, place, persons, and all circumstances apologize for me, and why may I not then be idle with others, speak my mind freely? If you deny me this liberty, upon these presumptions I will take it: I say again, I will take it.
>
> (p. 122)

And although he goes on here to "recant" and apologize, he has, already, "taken" his liberty, in his Utopian interlude, introduced with an assertion of authorship and individual authority:

> I will yet, to satisfy and please myself, make an Utopia of mine own, a New Atlantis, a poetical commonwealth of mine own, in which I will freely domineer, build cities, make laws, statutes, as I list myself. And why may I not? *Pictoribus atque poetis,* etc.—you know what liberty poets ever had, and besides, my predecessor Democritus was a politician, a recorder of Abdera, a law-maker, as some say; and why may not I presume so much as he did?
>
> (pp. 97–98)

The repeated assertion of personal authorship ("of mine own"), linked with the power of authority ("I will freely domineer"), is capped by the concept of poetic license *("Pictoribus atque poetis"),* which attributes a domain of free expression to the author. Yet all this is accompanied by recourse to the sanction of custom ("what liberty poets ever had") and, once again, to the authority-by-association of Democritus ("my predecessor"). Burton insists upon his rights as poet to make, control, and authorize his own vision of the world, while simultaneously attempting to legitimize and justify the endeavor with the sanction of convention and time-honored authority.

It may seem that I have wandered far from dwarfs and giants, but I have tried to suggest how the complicated use of that analogy throughout its history reflects the issues of authority and authorship. I have taken my two major examples from as early as the twelfth century and as late as the seventeenth because they span the period I will discuss in subsequent chapters. I have tried here as well to indicate that the question of the authority of authors is addressed not only through the comparison but also in various ways throughout these works that allude to it, and that the relation between predecessors and successors is but one manifestation of the general tension between the assertion and denial of authorship, between the attribution of authority to self and to other sources, that is the subject of this study.

II

The title I have chosen for this study is intended to convey the full dimensions of the subject, and it derives from the concept of poetic license alluded to in the last passage I quoted from Burton: "*Pictoribus atque poetis*, etc.—you know what liberty poets ever had." Burton's brief reference, as we saw, culminates an argument for the individual authority of authorship, yet the idea does not stand alone, for it both includes and generates an appeal to different forms of external, prior, and traditional authorization. Poetic license is historically a concept that invokes the various possible forms the relationship between authorial autonomy and authoritative sanction may take, and that delineates the theoretical considerations involved.

Sir John Harington, in his *Preface, or rather Briefe Apologie of Poetrie, and of the Author and Translator*, prefixed to his 1591 translation of *Orlando Furioso*, defends poets against the charge of lying first by calling upon this idea of poetic license:

> And first for lying, I might if I list excuse it by the rule of *Poetica licentia*, and claime a priviledge given to Poet[s], whose art is but an imitation (as *Aristotle* calleth it), & therefore are allowed to faine what they list, according to that old verse,
>
> *Iuridicis, Erebo, fisco, fas vivere* [r]*apto;*
> *Militibus, medicis, tortori, occidere ludo est;*
> *Mentiri astronomis, pictoribus atque poetis,*
>
> which, because I count it without reason, I will English without rime.
>
> Lawyers, Hell and the Checquer are allowed to live on spoile;
> Souldiers, Phisicians, and Hangmen make a sport of murther;
> Astronomers, Painters, and Poets may lye by authoritie.[24]

This is a far cry from Sidney's defense against the same charge. Although Harington next echoes Sidney's rebuttal that poets cannot be called liars because they never claim that their fictions are true, he initially presents the argument that lying is the privilege of poets: "Thus you see that Poets may lye if they list *Cum privelegio*" (p. 201).

This concept of poetic license that Harington employs in his rather weak justification of poetic feigning is complex. On the one hand, it implies autonomy, the personal and professional liberty of poets to disregard existing rules, standards, and conventions and, in Harington's words, "to faine what they list." On the other hand, it also suggests a kind of regulation: the granting of formal, official permission or approval for a specific activity, *by* rule or convention. Harington's own attempt to sanction poetic license "according to that old verse" implies this second meaning, as does his reference to the "rule" of poetic license as a right "given" to poets. We have only to recall the acts of the Privy Council and the Star Chamber, and the licensing orders of the sixteenth and seventeenth centuries, to comprehend the full impact of this latter notion of license. Poetic license is a concept that invokes, in other words, a sense of freedom from external laws as well as governance by them. To say then that poets "lye by authoritie" is to beg a crucial question: Harington conspicuously avoids specifying the source of authority. Does the poet say what he wants on his own authority, or by permission granted by some higher power?

The notion of authority attributed to the poet's license is complicated further by the passage that introduces the charge of lying to which Harington is here responding. Harington cites Cornelius Agrippa's *De Vanitate et Incertitudine Omnium Scientiarum* as the source of the chief objections to poetry, but he first describes Agrippa as "a man of learning & authoritie not to be despised" (p. 199). After summing up Agrippa's "bitter invective against Poets and Poesie," Harington begins to retract the original description of him, warning those who "wil urge this mans authoritie to the disgrace of Poetrie" to be careful that their actions do not backfire, because Agrippa attacks almost all classes and professions, including the court, the aristocrats, priests, lawyers, and physicians. He concludes that "I thinke it were a good motion . . . that this mans authoritie should be utterly adnihilated . . ." And yet he adds that, though he intends to refute Agrippa's objections against poetry, "but for the rejecting of his writings, I refer it to others that have powre to do it, and to condemne him for a generall libeller" (pp. 199–200).

What has happened to the "man of authoritie"? Within the space of one paragraph, his authority has been completely transformed into something worthy not of respect but of scorn and rejection. At the same time, Harington's admission that he does not have the "powre" to strip

Agrippa of his authority indicates that Agrippa still retains something of his early status; yet he also implies that there is a competing group of "others that have powre to do it." Authority, in other words, is here presented from various perspectives: as a tenuous quality that can be lost or denied, and also as a tenacious one, not easily challenged or destroyed; as an illegitimate and dubious possession of certain men who do not rightfully deserve it and who abuse it; as a characteristic of rival communities; and as something worthy potentially of respect as well as of condemnation. When Harington proceeds in the next paragraph to provide his own translation of "that old verse" that defines "the rule of *Poetica licentia*"—"Poets may lye by authoritie"—all these different dimensions of authority are recalled, and they serve to highlight the complexity of the concept of poetic license, which essentially invokes the presence of two systems of authority: the autonomous author with license (or freedom) in the poetic domain, exempt from external control and conventional rules, and the external power that bestows the license and hence authorizes the poet's actions.

The word *license* also has a third connotation that must be added to the idea of official approval and the positive implications of freedom from restraint. It is suggested when Harington correlates poets who "lye by authoritie" with the anarchic excesses of lawyers, for example, who "live on spoile," and physicians and hangmen who "make a sport of murther." Milton specifically invokes this third sense in *Areopagitica* (subtitled *A Speech for the Liberty of Unlicensed Printing*), when he writes that he should not be accused of "introducing license, while I oppose licensing."[25] In the name of "liberty" he opposes licensing, or official (legal and civil) control, but he also opposes license as the excessive exploitation or abuse of liberty. For Milton, the individual license of completely lawless freedom and the public license of official control are similar in their negative qualities.[26] When in Sonnet 12 ("I did but prompt the age") he writes of those who have misunderstood his divorce tracts, "License they mean when they cry liberty," he further implies that while licensing may curtail freedom, absolute license may do so as well, and that true freedom—as distinguished from both external regulation and unbridled lawlessness—carries with it a private set of responsibilities for the individual. In short, there is a need to mediate between the license of external constraint and the license of wholly uncontrolled freedom, each of which may be equally dangerous. Harington acknowledges this need as well when, after justifiying the poet's privilege to say what he list, even to lie, by the rule of poetic license, he proceeds to insist that poets "cannot lye though they would" (p. 201) because of the internal law that guides and defines their work: they never affirm their fables as truth. Thus, in addition to the competing claims of the poet's autonomy and external control, the idea of poetic license also invokes the possibility of reckless excess and implicitly suggests the inner constraints and personal responsibility that must accompany individual freedom.

The history of this notion of poetic license, in its many different forms, reveals a constant attempt to deliberate and negotiate among these various implications. The concept dates at least as far back as Horace, who is quoted by Burton and echoed in the last phrase of the verse that Harington cites: "*Pictoribus atque poetis . . .*" In the *Ars Poetica,* Horace himself cites the idea as if it were familiar in his time as well. It is the response he anticipates from his audience to his opening statement concerning the need for simplicity and unity in poetry: "pictoribus atque poetis quidlibet audendi semper fuit aequa potestas" ("Painters and poets have always had an equal right in hazarding anything [they please]").[27] This phrase is often quoted by Renaissance writers as testimony to the poet's freedom, yet Horace mentions it only so that he may temper it, for he proceeds to specify the particular rights and responsibilities of that "license":

> We know it: this license we poets claim and in our turn we grant the like; but not so far that savage should mate with tame, or serpents couple with birds, lambs with tigers.[28]

Horace, too, speaks for freedom within limits—for a license that is both claimed and granted and that follows certain rules (here, of probability and consistency).

A similar note is sounded by medieval grammarians and rhetoricians, who speak most frequently of a more narrow form of poetic license—one that Quintilian alludes to in his *Institutes* and that is associated primarily with diction, specifically with the *metaplasm*. The *metaplasm* was considered an acceptable modification of or irregularity in the form of a word in poetry; in prose, the general name for this was *barbarism*, which was deemed a flaw or fault. In his description of how the teacher of grammar or literature should analyze verse, Quintilian explains:

> He will point out what words are barbarous, what improperly used, and what are contrary to the laws of language. He will not do this by way of censuring the poets for such peculiarities, for poets are usually the servants of their metres, and are allowed such license that faults are given other names when they occur in poetry: for we style them *metaplasms, schematisms* and *schemata* . . . and make a virtue of necessity.[29]

Quintilian associates the poet's license with a freedom to deviate from the "laws of language," particularly because of the metrical demands of verse. Yet, clearly, this freedom is based on a premise of servitude (poets are "servants" of their meters) and "necessity," and hence is "allowed" rather than censured as vice or error. This "doctrine of the 'permitted fault' "[30] also appears in the fourth-century *Ars Major* of Donatus, who devotes an entire book of his work to *barbarisms* and explains that the *metaplasm* is a "change in a word for the sake of metrical ornament."[31] In the seventh century, Isidore of Seville, in his *Etymologia*, mentions and further develops this idea that poets have license to change the form of words and to depart from the norms of language usage; for him, *metaplasm* is the "transformation of a word on account of metrical necessity or poetic license" (*licentia poetarum*).[32] The concept now begins to expand into the category of figures. A twelfth-century formulation of this idea is found in Averroes's commentary on Aristotle's *Poetics*. According to Averroes, "variation from proper and standard speech" is a defining feature of poetry, achieved through accents, rhyme, metaphor, and "in general, through any sort of departure in language from standard usage . . . and through all the techniques that fall under the heading of poetic license."[33] Poetic license, by this time, though still associated primarily with changes in the forms of words, is more generally conceived of as a deliberate departure from conventional standards. It thus implies a broader concept of poetic freedom, but it sustains the idea of a governing authority through the almost constant reminder that such deviations are *permitted* and that approval is being granted for what in other contexts would be considered flaws. John of Salisbury, who in his *Metalogicon* quotes with approbation the technique of teaching described by Quintilian (I.24; p. 66), also refers his readers to Isidore of Seville to acquire the necessary "rules" that not only point out the "faults" of "grammatical errors to be avoided" but also clarify "the questions as to what forms of metaplasms, schemata, and tropes are permissible and ornamental" (I.20; pp. 59–60). Yet when he explains that "It is called a 'metaplasm,' or a sort of 'transformation' or 'deformation,' because, as though on its own authority [*quasi a iure suo*], it modifies or disfigures the form of words" (I.18; p. 53), he implies not simply that such deviations from the rules are externally legitimized but that they are self-justifying. John concludes by mediating between these viewpoints:

> License to use figures is reserved for authors and for those like them, namely the very learned. Such have understood why [and how] to use certain expressions and not use others. According to Cicero, "by their great and divine good writings they have merited this privilege," which they still enjoy. The authority of such persons is by no means slight, and if they have said or done something, this suffices to win praise for it, or [at least] to absolve it from stigma.
>
> (I.18; p. 54)

The license to deviate from the rules is "reserved for" or granted to (*licentia indulgetur*) only those authors who have proved their right to such a privilege by establishing their own authority, an authority that then suffices to validate whatever they do. This license, in short, must be earned; it both confers and acknowledges authority; and it provides a domain for the writer in which his own status serves to authorize what (and how) he writes.

We see then that the medieval concept of poetic license, along with its ancient antecedents, anticipates the Renaissance notion, which is frequently applied more broadly to a general idea of poetic freedom of speech, not limited merely to language usage. In fact, the medieval concept survives in the Renaissance and coexists with the expanded definition.[34] Both imply a similar tension between external authorization and individual authority and autonomy. As Gabriel Harvey writes, poetic license does not necessarily secure freedom, for it implicitly includes restrictions as well:

> Oratours have challenged a speciall Liberty: and Poets claimed an absolute Licence: but no Liberty without boundes: nor any Licence without limitation.[35]

Nor does it provide any guarantees of protection or privilege: "a Poets or Painters Licence, is a poore security, to privilege debt, or diffamacion."[36] Poetic license may be, as Gascoigne calls it, "a shrewde fellow," but it must tread carefully, for its privileges are as myriad as its constraints, its freedoms as precarious as the sanction it provides.

Bacon's brief discussion of poetry in *The Advancement of Learning* is instructive as it tackles this difficult issue, for he attempts to delineate separate areas of poetic limitation and liberty. According to Bacon, poetry "is taken in two senses, in respect of words or matter":

> Poesy is a part of learning in measure of words for the most part restrained, but in all other points extremely licensed, and doth truly refer to the Imagination; which, being not tied to the laws of matter, may at pleasure join that which nature hath severed, and sever that which nature hath joined, and so make unlawful matches and divorces of things: *Pictoribus atque poetis*, &c.[37]

With respect to words, poetry is "restrained"; with respect to matter, it is "licensed." Bacon's primary interest in this passage is with "matter," and he quotes the familiar Horatian line as testimony of the poet's extreme license to form "unlawful" creations "at pleasure." Despite his customary distrust of imaginative excesses, Bacon here ignores—as most writers of his time did—the specific qualifications of this concept that Horace immediately includes in his *Ars Poetica,* where he ultimately rejects a poet's privilege to make "unlawful matches." But Bacon nonetheless—again, as others did—adds his own. Although he claims that "style" (or words) is "not pertinent" to his present discussion of "matter," his earlier declaration that "words are but the images of matter"[38] makes his reference to them here very pertinent indeed. Like Harington, whose description of absolute liberty undercuts his apparent insistence on it, Bacon qualifies his statement concerning the poet's imaginative freedom by juxtaposing it to the restraints that the very medium of language imposes.

Poetic license, then, is never absolute. Throughout its early history it has always implied a competition between liberty and restraint, between the autonomy of authorship and its regulation and sanction by other sources, and it conjures up the various contingencies that both determine and deter a writer's privileges and responsibilities to authorize his own work.

III

In her penetrating analysis of authority (its origins and its virtual disappearance in the modern world), Hannah Arendt defines authority in political terms in a way consistent with the implications we have drawn from the phrase "poetic license."[39] As she explains, "authority always demands obedience," and yet the authoritarian order is "always hierarchical" and is itself "bound by laws."[40] She elaborates:

> The source of authority in authoritarian government is always a force external and superior to its own power; it is always this source, this external force which transcends the political realm, from which the authorities derive their "authority," that is, their legitimacy, and against which their power can be checked.[41]

In other words, Arendt sees true authority as derived from some external power or rules and, therefore, as subject to and restrained by them. Her model for this is the shape of the pyramid, a "fitting image," she explains, for a structure "whose seat of power is located at the top" but "whose source of authority lies outside of itself" and "above it."[42] Like the literary authority of authors implied in the notion of poetic license, Arendt's definition invokes a complex network of freedom and limitation, individual autonomy and external regulation, in which authority may be contingent, derivative, and ultimately answerable to (and measurable against) an outside and superior source.

In her essay, Arendt also discusses the etymology of the term *authority*, which she traces back to Roman origins. *Auctoritas*, she claims, derives from *augere* ("to augment"), and is opposed, according to her analysis, to the *artifex*, or "maker"; the true *auctor*, who has authority, neither makes nor builds, but "inspires." When she adds that despite its "binding force," the "most conspicuous characteristic of those in authority is that they do not have power," she admits that authority and the function of authors may seem, under these circumstances, "curiously elusive and intangible."[43] As Laurence Holland, who has discussed the bearing of Arendt's ideas on literary matters, comments, "in this etherealization of authority, Professor Arendt comes close to refining authority out of existence." He uncovers the ultimate consequence of a definition that gives the author such a "comparatively passive, ethereal function": "the perfection of authority would be silence."[44]

Arendt's study clearly has important implications for literary studies, and it raises crucial questions about the authority of the author as writer. If an author is seen as—or sees himself as—a maker, what then is his relation to authority? In literature, is there a distinction to be made between the *artifex*—the maker—and the authority—the one who "inspired the whole enterprise"? Or do literary makers attempt to close that gap, to claim their authorial authority *as* makers? Does the status of maker suggest a source of authority that presides externally or one that resides internally? Does silence signify the perfection of authority, or the complete loss of it for the author whose medium is language?

These are the questions this study addresses by examining the way a range of early writers have confronted them within a range of literary frameworks. As we prepare to turn now to these specific contexts, it is instructive to note how Arendt's etymological analysis of the relation between *auctoritas* and *auctor* compares to the way early writers used these terms. A. J. Minnis's discussion of this provides a detailed account:

> According to medieval grammarians, the term [*auctor*] derived its meaning from four main sources: *auctor* was supposed to be related to the Latin verbs *agere* 'to act or perform', *augere* 'to grow' and *auieo* 'to tie', and to the Greek noun *autentim* 'authority'. An *auctor* 'performed' the act of writing. He brought something into being, caused it to 'grow'. In the more specialised sense related to *auieo*, poets like Virgil and Lucan were *auctores* in that they had 'tied' together their verses with feet and metres. To the ideas of achievement and growth was easily assimilated the idea of authenticity or 'authoritativeness'.[45]

Yet in attempts to establish the status of the human *auctor* and the *auctoritas* of Scripture, significant distinctions were made: "Where there is a craftsman (*artifex*) who directs and guides a work and another who works with his hands in accordance with rules conveyed from the craftsman, the latter is not said to be the *auctor* of the work but rather the former."[46] Although ultimately, as Minnis explains, the divine *auctoritas* of Scripture did not interfere with the growing emphasis on the "integrity of the human *auctor*," it is clear that at stake was the relation between the *auctor* (both as inspirer and maker) of the text and the source of the text's authority. Minnis also notes that some writers made a distinction between *actores*, or "mere writers," and *auctores*, or "writers who are authorities."[47] In other words, despite the etymological association of *auctor* with the "performance" (*agere*) of writing (especially of poetry [*auieo*]), with the notion of inspiring growth (*augere*), and with authority (*autentim*), the idea of "authoritativeness" was only selectively and partially assimilated to this conception of the author and to "the act of writing."

A final helpful guide may be Dante, one author who provides his own etymology of the word *authority*. In his *Convivio*, Dante initially offers a rather simple definition: "Authority is no other than the act of the Author."[48] He then proceeds to discuss the "two roots" of the term *auctore*. The first is a Latin root:

> One is from a verb, whose use in grammar is much abandoned, which signifies to bind or to tie words together, that is, AUIEO . . . and how much *Autore* (Author) derives its origin from this word, one learns from the poets alone, who have bound their words together with the art of harmony.[49]

The second is Greek:

> The other root from which the word "Autore" (Author) is derived, as Uguccione testifies in the beginning of his Derivations, is a Greek work, "Autentim," which in Latin means "worthy of faith and obedience." And thus "Autore" (Author), derived from this, is taken for any person worthy to be believed and obeyed . . . Wherefore one can see that Authority is equivalent to an act worthy of faith and obedience.[50]

Dante's first derivation directly identifies the *auctore* with the poet and his words; his second defines authority as an act of any person worthy of being believed and obeyed. This would seem to conform to Minnis's general account that "the term *auctor* denoted someone who was at once a writer and an authority, someone not merely to be read but also to be respected and believed."[51] Taken together, Dante's two etymologies associate the poet with authority and make his language a constituent feature of that authority, as though the poet's use of language carries its own sanction, evoking credibility and respect. But Dante does not incorporate these two etymologies: he elaborates the first but implies that it is obsolete and concludes that it is not pertinent to his present discussion of the term *authority*; of the second he says, "and thence comes this word, of which one treats at the present moment, that is to say, Authority." Dante in this way suggests a connection between poets and authority but also marks a separation between them, leaving unresolved the crucial and vexed question of whether the words bound together harmoniously by poets are to be distinguished from or identified with the act of authority worthy of faith and belief.

CHAPTER II

Dream Visions of *Auctorite*

I

In the *Pastime of Pleasure* (1509), Stephen Hawes presented his evaluation of Chaucer:

> The boke of fame / whiche is sentencyous
> He drewe hymselfe / on his owne inuēcyon
> And than the tragydyes / so pytous
> Of the nyntene ladyes / was his traslacyon
> And vpon his ymagynacyon
> He made also / the tales of Caunterbury
> Some vertuous / and some glade and mery[1]

Hawes appears to be making some fine, if curious, distinctions. He ascribes the *House of Fame*—which has been called "among the most independently-imagined of Chaucer's narrative poems,"[2] and yet is cast into a form (the dream vision) that supports a distinct, ancient literary heritage—to the author's "owne inuēcyon." He cites the *Legend of Good Women* as a "traslacyon" while neglecting the most obvious works in Chaucer's canon that would more neatly fit this category—the *Consolation of Philosophy*, the *Roman de la Rose* (the *Legend* even refers to these as the narrator's translations). And the *Canterbury Tales*, whose narrator insists that he is faithfully transcribing the stories of other people, are simply labeled, without discussion, the product of "ymagynacyon." While Hawes here conspicuously refrains from acknowledging the complexity (both for the Renaissance and the Middle Ages) of the issues and terms he invokes, he in fact offers some crucial insights into them. Of particular interest in this regard is his use of the phrase "his owne inuēcyon" in relation to Chaucer's *House of Fame*. The dual meaning surrounding this phrase by the time the *Pastime of Pleasure* was published—"invention" as the selection of topics or arguments, and "invention" as original creation[3]—makes it an oddly appropriate commentary on this fourteenth-century dream vision: a poem in which the speaker's arrogantly independent stance of "non other auctour alegge I" is accompanied by his self-effacing and deferential gesture of "the book seyth."[4] Indeed, Hawes's seemingly facile identification of the origin of Chaucer's poems suggests a source of major concern for Chaucer and many of his contemporaries. For these men felt strongly the need to define the (often uneasy) relationship between the status of their individual authorial positions and the authorizing principles of their art—to find the proper balance between their claims for poetic independence and their reliance upon the sanction of traditional *auctorite*.

The medieval dream vision, in particular, can be considered a literary form especially suited to, and even generated by, the attempt to locate an authoritative perspective or interpretation with which the author may associate himself or from which he may speak—one that differs from his personal voice, which discovers, produces, and wrestles with the difficulties of the dream content. But the poems are also informed by a discovery of the unavailability—or inadequacy—of the conventional authorities appealed to and tested. Furthermore, a fundamental conflict exists between (1) the persisting desire to accept a traditional authority as ordering principle for the condition described in the poem and for the poem itself, and (2) the speaker's growing unwillingness to relinquish his own authority—his reluctance to engage in any betrayal of the authorial voice that, the poems suggest, the first act requires.

The nature of authority was an issue built into the dream vision, not only because its literary form generally included the appearance of an authoritative guide whose function was to teach some religious or secular doctrine or to explain the significance of the dreamer's experience,[5] but also because the status of dreams and dream interpretation was itself an unresolved matter of frequent deliberation. Cicero's discussion in *De Divinatione* reviews some major areas of concern in antiquity. Of primary importance was determining the validity and authenticity of dreams, and underlying the distinction between true and false dreams that was alternately upheld and contested stood the concept of a hierarchy that disparaged the role of the individual. Although Cicero's purpose, following Aristotle, is to oppose the notion that dreams may be sent from the heavens ("it must be understood that there is no divine power which creates dreams"[6]), he first presents the argument—a conventional and prevailing one—that dreams are to be identified by locating their origin: externally induced (i.e., divinely inspired) dreams are authentic and valuable; those produced by man himself (i.e., deriving from his personal thoughts and activities, his daily experience, his psychological and physical condition) are classified as meaningless and insignificant. This was also the attitude authorized by the biblical warning, ". . . neither hearken ye to your dreams which ye caused to be dreamed. For they prophesy falsely unto you in my name: I have not sent them, saith Jehovah" (Jer. 29:8–9).[7] The individual's participation in creating his dream thus immediately and necessarily invalidates it; the only sanction for a dream derives from a source extrinsic to the mind of the dreamer. And yet, despite this denial of the power of individual authority, a person could not totally be divorced from the significance of his dream, for general opinion acknowledged that a man's dream carries his personal stamp. It was an ancient idea that the social position and cultural rank of the dreamer helped determine the meaning of his dream;[8] aspects of personality and the condition of the mind were judged influential as well, and had to be considered when interpreting dreams. As Cicero states, "the same dreams have certain consequences for one person and different consequences for another and . . . even for the same individual the same dream is not always followed by the same result."[9] On one hand, the principles authorizing dreams excluded the possibility of any legitimate individual contribution, and on the other, it was not easy to dismiss the individual's unique role in the production of his dream. Complete, unqualified authority was difficult to claim for any source, whether external or internal to the dreamer. In *De Divinatione*, for example, Cicero did not extend his rejection of divine authorization and his acceptance of the personal factor, as he might have, to a conclusion that conferred authority on the individual dreamer; rather, from these premises he deduced the total absence of authorizing principles and concluded that "absolutely no reliance can be placed in dreams."[10] The endeavor to define a place for the individual complicated, and often made impossible, attempts to produce a general, consistent set of rules to govern dream interpretation.

Similar problems were presented to and by medieval writers.[11] The outlines of the issue in this period can be observed by setting together two of the most influential analyses of dream lore available in the Middle Ages, Macrobius's commentary on the *Dream of Scipio* and John of

Salisbury's discourse on dreams in his *Policraticus*. Macrobius formulates a science of dreams, providing guidelines for classification and interpretation. He differentiates dreams according to five categories, two of which (*insomnium* and *visum*) are unreliable, without importance or significance, and "not worth interpreting."[12] The other three are reliable dreams for which Macrobius seems to offer distinct definitions and explanations, but the divisions are not always helpful, and the criteria for each category suggest the dilemma that confronted the would-be practitioners of dream interpretation. One type, the prophetic dream (*visio*), can be so identified only after experience corroborates or belies it: "We call a dream a prophetic vision if it actually comes true." Nothing intrinsic to the dream determines its legitimacy. The two remaining types of authentic visions are determined by authoritative explanations that either are delivered within the dream or are imposed from outside. In the oracular dream (*oraculum*), a "parent, or a pious or revered man, or a priest, or even a god clearly reveals what will or will not transpire, and what action to take or to avoid." For the enigmatic dream (*somnium*), the authoritative interpretation is not internal; this type "conceals with strange shapes and veils with ambiguity the true meaning of the information being offered, and requires an interpretation for its understanding."[13]

Macrobius's systematic description of the dream phenomenon exposes some critical issues. There is, first of all, a need to differentiate the reliable and significant (i.e., prophetic) dream from the insignificant (physically or psychologically induced). In one case, however, this distinction can only be made after the fact, when prophetic information is no longer valuable because it has already been realized. In the other cases, the authority that interprets from either inside or outside the dream becomes the vital feature, yet the identifying characteristics of this figure are, as presented, extremely broad as well as various, and there is no mention of the interpretive rules that govern his activity. Macrobius does not acknowledge any problems or omissions in his outline of dream theory, but his syllabus left room for resistance from those who were unwilling to accept the validity of the criteria he proposed—and even of the entire endeavor he was pursuing.

John of Salisbury was one such acute dissenter who could not accept Macrobius's word as final or complete and who ultimately came to reject the art of dream interpretation. Although John does not speak directly to the issue of *literary* dreams in his *Policraticus*, his allusions to the Macrobian system provide a useful commentary on some of its troublesome areas. John reiterates Macrobius's classifications with apparent sanction, but as he proceeds to add his own commentary on them, he complicates the crucial matters so that even the "reliable" dreams become suspect. The figure of authority in the oracular dream, for example, is carefully dissected, and he emerges in an extremely precarious position. John at first identifies this man by echoing Macrobius: he acquires his status as an honorable man worthy of reverence "either as the result of nature as in the case of a parent, of position as in the case of a master, of character as in a man of piety, of chance as in a magistrate, of religion as in a god, angel, or man consecrated to holy office."[14] But John goes on to discuss the implications of these broad defining features: "From this it is apparent, if not directly yet by inference, that in respect to the art of interpreting dreams individuals not merely lacking honor but even accursed are included in the class worthy of reverence" (II.15; p. 79). The reliability of the authoritative figure—the defining character in Macrobius's system—is thereby called into question, and the issue is no longer merely that of distinguishing the true dream from the false: now the authority that simultaneously signifies a valid dream and interprets its meaning becomes subject to duplicity. In John's system, this figure is implicated as part of the dilemma for which Macrobius considered him a solution.

The external interpreter who, according to Macrobius, explains the meaning of enigmatic dreams, also acquires a more complicated function from John's perspective. Interpretation itself becomes an extremely intricate activity, because "any particular thing has inherent in it as many meanings of other objects as it has likenesses to them . . ." (II.16; p. 81). Since on the one hand, "signs are frequently the same" (II.16; p. 81), and on the other, "All things involve varied and manifold meanings" (II.17; p. 84), the dream interpreter must be careful not to pursue doggedly only one line of interpretation and consequently declare one particular meaning when there are so many other available meanings to consider. By this point in his discussion, John seem to disallow the entire practice of interpretation by denying the interpreter a secure place to rest; the multiple meanings of "signs" defer the possibility of choosing a specific explication and granting it a privileged status. Indeed, as John himself finally contends,

> Hence the dream interpreter which is inscribed with the name of Daniel is apparently lacking in the weight which truth carries, when it allows but one meaning to one thing. This matter really needs no further consideration since the whole tradition of this activity is foolish and the circulating manual of dream interpretation passes brazenly from hand to hand of the curious.
>
> (II.17; p. 84)

Thus, although Macrobius provided the Middle Ages with guidelines for dream interpretation, setting up certain features as characteristics of authentic, authoritative visions, dream theory was also derived from men like

John of Salisbury who did not consider these classifications dependable. John's commentary invades the domain of the "true" dreams to invalidate both the authoritative figure who appears within the dream and the one (the interpreter) who functions outside of it. He presents Macrobius's categories and then begins to obscure the distinctions between them, so that the structure of dream theory finally collapses into a muddle of indistinguishable and ambiguous criteria that denies the validity of the entire exercise of dream interpretation. In his final chapter on dreams, John concludes that dream interpretation is not to be recommended since it is "no art or at best a meaningless one" (II.17; p. 84). Some select few may, on occasion, be visited by divine illumination that will explain a dream, but to turn this into an art based on human experience and reason, to deposit one's faith in dreams and dream interpretation, is to embrace deception and infidelity.

Medieval dream theory, then, centered around questions of authority and authorship. False and deceptive dreams had to be distinguished from authentic and prophetic dreams. To establish the existence of a true and significant dream, one had to locate first the origin of the vision and then either a figure within it whose very appearance would sanction the dream or one outside of it whose interpretation of its meaning was dependable. The question was whether the source could be clearly discerned, and, if so, whether there was any authoritative figure who could be unambiguously identified and whose reading could be considered reliable or complete.

The issues discussed by the prose commentators were echoed in many literary texts, whose authors confronted the problem not only of determining the authenticity of a particular dream, but also of choosing an authority among the dream theorists. One of the most influential and popular dream visions of the thirteenth century develops the issue in precisely these terms. Guillaume de Lorris opens the *Roman de la Rose* with what later becomes a conventional pattern: an acknowledgment that dreams may be true or false, followed by an assertion of the validity of the particular dream about to be reported:[15]

> Many a man holds dreams to be but lies,
> All fabulous; but there have been some dreams
> No whit deceptive, as was later found.
> Well might one cite Macrobius, who wrote
> The story of the Dream of Scipio,
> And was assured that dreams are ofttimes true.
> ....
> Now, as for me, I have full confidence
> That visions are significant to man
> Of good and evil. Many dream at night
> Obscure forecasts of imminent events.[16]

The citing of Macrobius as an authority to uphold the authenticity of dreams elicits the narrator's confidence in their reliable and prophetic nature—an attitude he extends to include his own dream. As Sheila Delany notes, this claim is part of the larger and familiar convention prevalent in medieval literature that attempts to justify the literary work by referring to and inventing sources or asserting truth and verisimilitude.[17] Recourse to historical (or pseudohistorical) support to authorize a poetic, fictional endeavor that otherwise, sanctioned only by the individual authorial vision and voice, could not be left to stand alone, suggests in broader terms the tensions that informed the use of the dream frame. The dream vision could become a vehicle for truth and revelation, if the proper authority was cited or the proper examples (mostly biblical) were presented. This "documentation" validated an expression that, by itself, could carry no claims to authenticity other than the (insufficient) personal testimony of its author or narrator. And yet, confronted with the complex matter of identifying a reliable dream and with frequent exhortations to abstain from such a dangerous, perhaps impossible task, this personal testimony came to stand as the only certification of the dream. When Jean de Meun deals with dreams in his section of the *Roman de la Rose*, he presents the view that made it difficult to accept the simple faith that Guillaume de Lorris's narrator voices at the beginning of the poem. It is no longer easy to side with Macrobius and thereby claim, with the support of an authoritative tradition, the legitimacy and significance of dreams (Jean places the *Dream of Scipio* in a category with lying and deceit), and the remaining option is to refrain altogether from making any judgments based on those principles:

> I will not say that dreams are true or false,
> Or whether men should all reject or none,
> Or why, when one may be most horrible,
> Another's most agreeable and fair,
> According as the apparitions come
> In various complexions of the mind
> Resulting from a difference of age
> Or habit; nor shall I attempt to say
> If God sends revelations by such dreams,
> Or if the evil spirits by them try
> To tempt men to their peril. None of this
> Will I discuss . . .[18]

This attitude is spoken by the Goddess Nature—a goddess whose stature as an authoritative figure has decreased considerably since her appearance in Bernard Silvestris's *De Mundi Universitate* and Alain de Lille's *De Planctu Naturae*. There she was depicted as a venerable, dignified ruling power (with acknowledged limitations); in the *Roman*, she is divested of her status and moved closer to earth, representing more the actual disruptive conditions of human existence than an ideal standard and governing force.[19] As the poem undermines this conventionally authoritative figure, it also undermines the authority of dreams as asserted at the beginning of the poem. The movement from the position of Guillaume de Lorris to that of Jean de Meun indicates that the supportive gesture of referring to traditional sources (e.g., Macrobius) is no longer helpful; no viable claims of authenticity or validity can be based on these conventional standards and criteria.

Presenting the opposing and contradictory views of dream visions became itself a convention, calling into question not only the dream but also the status and value of the *auctores* cited in support of either position. *Piers Plowman,* a dream vision popular well into the fifteenth and sixteenth centuries, is typical in this respect. Its narrator-dreamer conducts a quest for some authoritative standard or figure whose answers, interpretations, and solutions are both reliable and final. However, no such authority is discovered in the dream; various figures, representing both internal and external principles (Holy Church, Conscience, Reason, Scripture, Imagination, even Piers) are appealed to and tested in this role, but all fail to satisfy the dreamer, whose search confronts the disparity between the life he must live on earth, in practical terms, and the abstract, conventional formulations that are not accessible enough or adequate to accommodate each man's experience in the world.[20] As Morton Bloomfield has noted, "The poem deals with the search not the finding," and more recently Elizabeth Kirk has pointed out that the reader shares the narrator's unsuccessful quest: "We seek continually in the argument for a stability and satisfaction it does not provide."[21] This search for and failure to find an authoritative guide is an inherent feature of the form the dream vision had acquired by the fourteenth and fifteenth centuries, when it gained extreme popularity as a literary mode. Langland's characteristic technique of pitting text against text with contradictory, unresolved results, thereby casting doubt upon the authority of both sources, is employed effectively in his discussion of dreams. After the narrator has witnessed the "pardon scene," a crucial but inconclusive incident in which interpretation itself becomes the battleground, he is temporarily awakened by the dispute, and puzzles over the significance of this episode in particular and of his entire dream in general. He reviews the attitudes of various authorities but discovers from them no final answer:

> Many tyme this meteles · hath maked me to studye
> Of that I seigh slepyng · if it so be myȝte,
>
> ....
>
> Ac I haue no sauoure in songewarie · for I se it ofte faille;
> Catoun and cononistres · conseilleth vs to leue
> To sette sadnesse in songwarie · for, *sompnia ne cures.*
> Ac for the boke bible · bereth witnesse,
> How Danyel deuyned · the dremes of a kynge,
> That was Nabugodonosor · nempned of clerkis.
>
> ....
>
> And as Danyel deuyned · it dede it felle after,
>
> ....
>
> And Ioseph mette merueillously · how the mone and the sonne,
> And the elleuene sterres · hailsed hym alle.
> Thanne Iacob iugged · Iosephes sweuene:
>
> ....
>
> It bifel as his fader seyde . . .[22]

The conventionally reliable sources offer contradictory advice and evidence and therefore fail to provide helpful guidance. It then becomes the individual's business to either choose from the authorities one attitude to call his own, reject the traditions and rely on his own judgment, or abstain from taking any position at all. In this scene, the dreamer, after debating with himself, interprets his dream and then goes off on a new quest, seeking an authoritative identification, explanation, and sanction of the terms of his own interpretive choice.

These variable conceptions of dreams and of the value—even the possibility—of dream interpretation cannot be discounted when we consider the function of the dream frame as a literary device in medieval works. There was, to be sure, a heritage from classical literature that made the dream an appropriate vehicle for the teaching of secular or religious doctrine.[23] But some modern critics provide a one-sided picture when they ignore—or dismiss—the controversial currents of thought that would have surrounded the medieval author who chose to write in this tradition.[24] It cannot simply be assumed that the dream convention would immediately signify authoritative sanction; rather, it is likely that the choice of this framing device would indicate on the author's part, and perhaps invoke in his audience, an awareness of the very unsolved, ambiguous, and unstable nature of dreams, of interpretation, and of authority. And as the traditional supports for his work are undermined, the writer's position in relation to his text would become a prominent issue. If the dream frame no longer conjures up intimations of authority, then, as Delany notes, "neither dreaming nor writing offers any security of role" for the narrator.[25] Moreover, when the conflicting attitudes toward dreams are described and contained within the dream vision itself, calling into question the status of dream and dreamer, text and narrator, then the poem no longer carries its own justification and must seek to establish its legitimacy, perhaps on new grounds. The connection between writing and dreaming has been suggested often; J. A. W. Bennett points out that dreams were "traditionally associated with poetic creation,"[26] and other critics have recently claimed more generally that the unstructured, often associative nature of dreams encouraged the author employing one as a frame to give his imagination free play, since ordinary standards of credibility could not be applied to it.[27] This identification of the domain of dreams with that of literary invention makes the narrator as dreamer and poet an intrinsic part of his poem; he and his craft become, in some ways, the subject that is being explored.[28] For the dream frame may liberate, but when the creative act is granted a separate province that need not be accountable to conventional criteria, it also acquires sole responsibility for the status of the literary work. If dream visions allow for the *possibility* of freeing the poetic imagination, they certainly do not guarantee its acceptance or validity; claiming, demonstrating, and maintaining the separate and inherent rights of the creative mind and its products become the obligations, not merely the privileges, of the author. The "I" who dreams and writes may possess the power to certify his own vision, but this right would need to be established rather than assumed. Not only does Morton Bloomfield ignore the unsettled state of dream lore when he calls the dream frame an "authenticating device" that "served to suspend disbelief and to obtain credence"; his idea of the immediately self-validating impact of the dream vision simply assumes the authority of the writer ("A man telling his own dream usually tells the truth") and fails to account for the burden of proof that would fall upon him.[29] A. C. Spearing offers a significant qualification: ". . . the medieval reader was not necessarily called on to suspend his disbelief. The use of the dream-framework is frequently to evade the whole question of authenticity, of belief or disbelief."[30] I would go even further, however, and add that the evasion itself calls attention to—rather than dismisses—the problem of authorizing the work and brings the focus back to the authority of the "I." When the conventional standards of order and truth are rejected, a space is opened in which the individual vision or voice can emerge, however unqualified it may be, to be tested as its own principle of authority.

II

The dreamer-poet's journey to discover "newe tydynges"[31] in the *House of Fame* suggests an effort to locate the source of material—the origin—of his art.[32] The narrator is not, of course, Chaucer, but neither is he the terribly naive person who cannot comprehend the Black Knight's plight in the *Book of the Duchess*, not the man who, for the most part, simply observes and routinely records in the *Parliament of Fowls*. The overt references to personal matters (the narrator's name is "Geffrey," his occupation is discussed in a passage generally acknowledged as a description of the Controller of Customs [ll. 652–60], and he is a poet) indicate an obvious connection between narrator and author. But he is, on occasion, excessively frightened and hysterical (when he finds himself in the barren desert) or befuddled and dumbfounded (during the greater part of the flight with the eagle)—in other words, he is too unsophisticated, too unaware of the significance of his behavior to convince us that there is no distance conspicuously and deliberately drawn between himself and his creator. Sometimes, during long descriptive sections (for example, the retinue at Fame's court), he recedes and almost disappears both as character and as narrator, but he also at times comes forth with confidence to put the entire dream in perspective, asserting and defining his vision with a control that bespeaks authorial privilege and responsibility. He draws his function and identity from—and hence incorporates—two worlds whose distinctions he blurs: he is both a creature of and a creator of a literary fiction. I do not mean to imply that at specific points we may say that the narrator either represents or does not represent Chaucer, but rather that the shifting character and situation of the dreamer-poet—his naiveté, his exaggerated, comic responses, his solemn assurance, his perceptive eye, his disappearances, his sure-handed direction and mastery—create a context that allows Chaucer to explore, in its various aspects, the nature of the poet's position in relation to his text.[33]

The discussion of dreams that opens the *House of Fame* simultaneously adopts and departs from the conventional apparatus that I have already examined in relation to the *Roman de la Rose*. As in the earlier dream vision, the narrator records the problematic history of dream lore: its various (and often indistinct) categories and causes, and the opposing responses accordingly elicited concerning dream interpretation. Guillaume de Lorris's narrator reacts to his simplified outline of the controversy by calling in Macrobius as an authority with whom he confidently aligns himself and whom he then uses to support his claim that dreams are valid. Jean de Meun's section recites more particularly the details of the dispute and develops the final stance that no stance at all should be taken concerning the nature of dreams. These precedents suggest two alternative ways of dealing with an ambiguous tradition that engenders confusion and encourages contradictory attitudes: either to select and embrace one authority, ignoring its compromised position, or to conclude that since the conventional accounts offer no privileged, authoritative answer, nothing whatsoever can or should be said on the subject. In the *House of Fame*, the narrator's response takes a different direction. The effort to discover the source of poetry requires that the status of the author be accounted for, and when the conventional and external models of authority that sanction authorship of the dream vision are found inadequate, the speaker does not merely grasp one and disappear behind its reputation or give up and refuse to confront the issue. Instead, he gradually begins to assert his own voice and comes forth to audition for the role of authorizing principle that has been left unfilled.

The comprehensive survey of dream theory at the beginning of the poem reveals the speaker's full knowledge of the subject, while at the same time he declares, with professed simplicity, his inability to deal with it. He claims that he understands neither the baffling, incoherent rules by which dreams are categorized (ll. 7–12) nor the differentiation and significance of their various causes. During his fast-paced summary, he tries out several different ways of removing himself from the need to confront all this confusing material by handing over the problem—and the possibility of resolution—to others: leaving it, with a prayer, in God's hands (1, 57-58);[34] letting the "grete clerkys" work it out (53–54); in fact, letting *anyone* who can determine these things do so ("but whoso of these miracles / The causes knoweth bet than I, / Devyne he" [12–14]). The attitude at one point seems to echo the disavowal of participation that is deployed in Jean de Meun's *Roman*:

> For I of noon opinion
> Nyl as now make mensyon . . .
>
> (55–56)

But his removal of himself from the arena of discussion is finally as quickly abandoned as were the various attempts to attribute practicable answers to other parties. After dispensing with the traditional criteria used to determine the authenticity of dreams, the narrator concludes the proem by offering, as the only immediate evidence and support for his dream, his personal testimony: he had the dream, and no man could have had one as wonderful:

> For never, sith that I was born,
> Ne no man elles me beforn,
> Mette, I trowe stedfastly,
> So wonderful a drem as I
> The tenthe day now of Decembre,
> The which, as I kan now remembre,
> I wol yow tellen everydel.
>
> (59–65)

The claim for the uniqueness of his dream detaches it even more from the generalized—and controversial—dictates of the traditions of dream lore that have previously been rehearsed, and in such a context, the statement functions as an assertion of individuality and independence.[35] The speaker also alienates himself from convention by setting the dream in December (traditionally, dream visions occur in May) and by emphasizing the subjectivity of his account ("as I kan now remembre"). In this way, he draws attention to his own role in and influence on what we are now encouraged to view as a literary representation, and the suggestion of potential unreliability is paired with the promotion of the speaker as the only authorizing principle that can, with any justification, be called upon.

This stance is slightly retracted in the following invocation, when the speaker himself calls upon another authorizing principle. As part of this conventional appeal "Unto the god of slep" (69), the narrator asks for outside guidance to aid him in meeting a standard of correctness apparently measured by more than simply what "I kan now remembre": "Prey I that he wol me spede / My sweven for to telle *aryght*" (78–79; my emphasis). The next line, however, qualifies the authority of the figure appealed to with a conditional statement that implicitly questions his status and control: "Yf every drem stonde in his myght" (80). If this challenges the position of the god of sleep, it also challenges the position of the poet, who has requested this dubious guidance to "telle aryght" his story. Thus compromised, the poet shifts his invocation to God, "he that mover ys of al," with a simplified notion of divine justice that seeks to limit radically the interpretive freedom of all those to whom, in the proem, the narrator had relinquished full authority for analyzing the dream according to dream theory. Those who "mysdemen" the dream are to be visited by "every harm that any man / Hath had" (99–100), while those who "take hit wel" (91) are to have "joye" and "grace" and other bounties. The rules presented by the speaker are quite straightforward and completely personal: if you like his poem, you've done well; if not, you haven't. Yet, while he offers detailed explanations of the ways to "mysdemen" or *mis*judge ("thorgh malicious entencion," or "thorgh presumpcion, / Or hate, or skorn, or thorgh envye, / Dispit, or jape, or vilanye" [93-96]), he outlines only the barest, vaguest instructions to evoke proper judgment and acceptance ("That take hit wel and skorne hyt noght" [91]). Although the audience is seemingly invited to judge, it is actually being *commanded* to judge *favorably*. Such almost tyrannical dictates, comically presented but poorly masking anxiety, may be necessary because, *pace* Bloomfield, this dream vision could not expect to obtain credence automatically by association with traditional supports, or by the standards of conventional interpretive guidelines. With only his own authority to back him up in a domain where the old identifiable rules do not apply, the narrator seems impelled to threaten (rather than convince) the members of his audience in order to win their belief in him and his story. The poem will continue to waver between the two alternatives sketched in the proem and invocation: reliance on external authorities that fail to maintain their positions and cannot accommodate the voice and vision of the speaker; or reliance on his own individual authority, which is never enough to control, explain, and interpret itself or the world that challenges it.

When the dream proper begins, the narrator is in the Temple of Venus, on one wall of which he finds the story of Dido and Aeneas, which he proceeds to recount at length. It is presented as a tale about both the betrayal of and fidelity to the different codes of behavior authorized respectively by two other records of the incident:

Ovid's interpretation of Aeneas as a traitor to his pledge of love to Dido, and Virgil's account of his heroic character and his allegiance to the gods who "bad hym goo" (430).[36] The greater part is devoted to the details of Aeneas's desertion of Dido and the nature (reliability) of the "trouthe" that is thereby compromised. Dido's mistake, we are told, was that she made of the untrustworthy Aeneas "hyr lyf, hir love, hir lust, hir lord" (258)—that she revered and "demed" good a man who "wolde hir of trouthe fayle" (297). This multifaceted "truth" is referred to its verbal representation, with which, we learn, it need bear no relationship:

> O, have ye men such godlyhede
> In speche, and never a del of trouthe?
> (331–32)

If a "trouthe" that does not "fayle" is to be discovered, it may not be in language. Neither words themselves nor those who speak them carry inherent authority, despite their seeming weight and power.

The problem for Dido—that "trouthe" may be betrayed by those who command belief by "godlyhede in speche"—is also a crucial narrative issue for the speaker and the story he is telling. In fact, two stories are told (creating a curious split in the narrative): after the hero's valiant exploits prior to his arrival in Carthage are related, the Ovidian attitude is evoked, which supports Dido's condemnation of a rather diminished Aeneas; then there is a sudden shift that works to "excusen Eneas / Fullyche" (427–28) by returning to the Virgilian perspective. A context similar to the one encountered in the proem's discussion of dream theory is thus created. Here the contradictory interpretations generated by the authority of literary history are exposed; the conventional materials fail to provide a satisfactory model and instead allow opposing alternatives to exist simultaneously. But it is more than the *existence* of this "dual tradition"[37] that makes the Dido and Aeneas story relevant: its significance to the *House of Fame* may also reside in the particular differences in attitude that Virgil and Ovid represent. George Kennedy expresses the distinction well in his description of the "disagreement between Plato and the sophists over rhetoric," which, he claims, "was not simply an historical contingency" but can be likened to the differences between a number of other figures, including Virgil and Ovid. Theses differences, he explains, reflect a timeless and "fundamental cleavage between two irreconcilable ways of viewing the world":

> There have always been those . . . who have emphasized goals and absolute standards and have talked much about truth, while there have been as many others to whom these concepts seem shadowy or imaginary and who find the only certain reality in the process of life and the present moment. . . . The difference is not only that between Plato and Gorgias, but between Demosthenes and Isocrates, *Virgil and Ovid*, Dante and Petrarch, and perhaps Milton and Shakespeare.[38]

Virgil and Ovid do not simply deliver contradictory "truths" concerning the Trojan story; their attitudes toward truth itself are entirely disparate. Their appearance in the *House of Fame* suggests a primary concern of the speaker in the poem: are there "absolute standards" that exist as authoritative, or is the only authority available something more local, relative, and individual? It is this latter view that can conceive of art as, to use Kennedy's phrase, "unconstrained by external principles." The debate posed by the speaker's literary predecessors exemplifies his problem as he attempts to locate and define his relation to an authoritative and authorial position. It may be seen as symbolic of the tension between his effort to discover and maintain allegiance to a reliable external source of authority and the opposite endeavor to stand by and authorize his own individual creation of literary truth. In fact, his oscillation between the Virgilian and Ovidian accounts mirrors the curious, more subtle, yet similarly shifting quality of his narrative stance during the retelling of the Dido and Aeneas story. The narrator tells us, at the beginning of this section, that the story he found on the wall of the Temple of Venus was

> Thus writen on a table of bras:
> "I wol now singen, yif I kan,
> The armes, and also the man
> That first cam, thurgh his destinee,
> Fugityf of Troy contree,
> In Itayle, with ful moche pyne
> Unto the strondes of Lavyne."
> (142–48)

These first six lines, set off in quotation marks, are the only words treated unambiguously as a faithful transcription of what we are told is the text appearing in the Temple.[39] For the next hundred lines, the narrator speaks as though he were paraphrasing, and his constant referrals back to the wall engravings ("ther sawgh I graven," "ther saugh I," and other variations: see 151, 162, 174, 193, 198, 209, 212, 221) draw attention to—and also indicate the speaker's preoccupation with—the artistic origin of the story he recounts, pulling us back from the action and plot to recall the mediation of the observer-transcriber and the text he claims as his source.[40] These frequent interruptions force us to be conscious of the mode and technique of narration as well as of the Trojan legend itself, and once this patterned refrain has been established, its absence becomes conspicuous. The speaker's stance as a recording agent, compelled simply to point out what is contained in his source, is not one that he maintains consistently throughout this section. Soon he seems to grow impatient with the demands of this strict allegiance; he not only ceases to acknowledge the text on the wall but also begins to assert his own voice and to assume certain authorial privileges. His primary concern is to trim and proportion the story to fit his own purposes and impulses:

> And, shortly of this thyng to pace . . .
> (239)

> That, shortly for to tellen . . .
> (242)

The speaker usurps the act of narrative composition and refers his text back not to what "saugh I graven" but rather to his personal faculties, talents, and predilections—in short, to his own artistic capabilities and preferences:

> What shulde I speke more queynte,
> Or peyne me my wordes peynte
> To speke of love? Hyt wol not be;
> I kan not of that faculte.
> (245–48)

His individual perspective and personality emerge to modify—even shape—the story he is retelling. He commands some power of judgment over his narrative and continues to indulge his authorial right to make technical decisions for himself, and now also for his audience:

> Hyt were a long proces to telle,
> And over-long for yow to dwelle.
> (251–52)

The speaker's intrusion as narrative designer is momentarily dropped as he returns to the familiar refrain of "ther sawgh I grave" (253, also 256), but despite the verbal recapitulation, he does not fully remove the evidence of his influence on and control over the text. After clearly and repeatedly establishing his desire to compress an "over-long" tale, he turns back to the wall of the Temple and there finds that:

> grave was, how shee [Dido]
> Made of hym *shortly at oo word*
> Hyr lyf, hir love, hir lust, hir lord . . .
> (256–58; my emphasis)

Here the character, although reinstated in the original text, nevertheless reflects and obliges the authorial impulses of the speaker. The dreamer-poet does not completely fulfill his formal gesture of reacknowledging a primary and controlling text other than his own: he resists a total surrender of his power to shape his narrative.

These rather quiet maneuvers around various authorial stances and attitudes are suddenly made more forceful and crucial by the events of the story itself. The pledge of love is followed by the account of Aeneas's desertion of Dido, during which no external source is mentioned. It is, in fact, presented in the form of the speaker's outburst: he moralizes (265–66), exclaims (265, 268), offers advice (269–70), refers to his personal situation (273), openly identifies the preceding statements as his own ("Al this seye I be Eneas / And Dido" [286–87]), and then announces and executes a conspicuously deliberate artistic choice:

> Therefore I wol seye a proverbe,
> That "he that fully knoweth th'erbe
> May saufly leye hyt to his ye;"
> Withoute drede, this ys no lye.
> (289–92)

The planning evident in the first line quoted indicates a narrative self-consciousness that does not appear during the preceding exclamations, and along with it arrives a complicating quality of caution and ambivalence. This passage intricately shifts among several different dimensions of authorial function and stature: it holds up the speaker as the controlling principle of narrative construction, but only so that he can incorporate a "proverbe"—a statement defined by the fact that it is not original with the author, that it has been previously ascertained and proved valid by common and familiar usage[41]—a proverb that he then, nevertheless, attempts to bring into his own domain with the weak flourish of his simple personal testimony, "this ys no lye." The wavering and ambivalence that characterize the narrator's efforts to assume full authorial rights over the text also restrain and betray that effort; moreover, they indicate that such a position is as untenable as subservience to an outside source seems to be.

The narrator acquires more strength as the story of the betrayal is drawn out, and his confidence in his own status and vision is heightened as he introduces Dido's complaint:

> In suche wordes gan to pleyne
> Dydo of hir grete peyne,
> As me mette redely;
> Non other auctour alegge I.
> (311–14)

The reference to the dream ("As me mette redely") reestablishes the basic context of the poem. More particularly, it recalls the proem, where the narrator affirms his right and responsibility to bypass the highly questioned rubrics of traditional dream theory and to deliver the final pronouncement concerning the legitimacy of his very individual dream. The following line ("Non other auctour alegge I"), clarifying this idea, more explicitly and firmly presents the narrator as authorizing his own text and is the most forceful statement we have had yet of authorial independence.

The earlier and briefer attempt to make narrative decisions (of proportion and structure) and to discard the refrain of subservience to the material on the Temple wall, begins at the point in the story that marks the departure from Virgil's account of Aeneas.[42] This second, more prolonged emergence of the narrator's authorial voice comes, significantly, when the story, revealing the influence of Ovid, describes and condemns Aeneas's departure as an act of betrayal and turns the pious hero into a treacherous womanizer.[43] On one level we can see that this episode in the story, which depicts the unreliability of words that carry with them "never a del of trouthe"—in fact, the potential unreliability of "trouthe" itself—coincides with the narrator's abandonment of his vows of fidelity to the Temple wall. In their place he substitutes his own voice, citing himself as the only source of the story he is telling. For if truth is compromised—particularly by the very language that purports to represent it—then the narrator can claim no certain status for himself or his poem by developing an allegiance (like Dido's) to a text or context that may (only) *appear* to be worthy and reliable. On another level, the introduction of the Ovidian material exposes the disparity between (and therefore questions the status of) two traditional, authorized accounts by turning the story around and creating a dishonorable Aeneas who directly contradicts the honorable one; and it is as this complication develops that the narrator, despite his echoing of Ovid's perspective, ceases to call upon the support of any other source. He proclaims instead an ethic of self-reliance. Rather than justify his tale by resting upon his status as a spokesman for what seems to be a previously authorized truth, he instead asserts complete autonomy, proposing that the full responsibility and power to validate his narrative reside in the exercise of his own authorial voice: "non other auctour alegge I."

As in the earlier incident of conspicuously displayed authorship, a retraction soon follows. Dido's complaint against the absence of "trouthe"—a complaint in which the speaker has already implicated his voice—is judged ineffectual: it "avayleth hir not a stre" (363). After noting this failure, the narrator gradually, tentatively descends from his lofty and isolated stance and begins to make careful references to some source material. He directs his audience to "Virgile in Eneydos / Or the Epistle of Ovyde" (378–79) to find the conclusion of the Dido story, which is "to long" for him to "here write" (381–82). A list of other betrayed women follows, taken from Ovid's *Heroides*, and here acknowledged as borrowed: "as the story telleth us" (406). As the catalogue of "untrouthes" grows uncomfortably large, however, the poet suddenly halts, switches to (though without naming) the Virgilian context, and reinterprets—so as to explain away—the major act of betrayal:

> But to excusen Eneas
> Fullyche of al his grete trespas,
> The book seyth Mercurie, sauns fayle,
> Bad hym goo into Itayle,
> And leve Auffrikes regioun,
> And Dido and hir faire toun.
>
> (427–32)

This "excuse" undermines the narrator's stance in the preceding two hundred lines (his claim of authorship as well as the account he authorizes), but it also eliminates the need for it. That peculiarly medieval gesture, "the book seyth," works to sanction Aeneas's desertion and, in fact, to redefine it as fidelity to a higher authority; "trouthe" is thus salvaged and reaffirmed. This benefit of the justification overwhelms the obvious element of contradiction, which receives no more attention than the "but" that introduces it. The return to Virgil's perspective (for which he reaches beyond the Temple to "the book") coincides with the narrator's return to the text on the wall: without any acknowledgment of his own "departure," he suddenly reinstitutes the refrain of "sawgh I grave" (433, 439, 451), and for the remainder of the tale of Aeneas's exploits he dutifully enacts his original pose as transcriber. The narrator returns to embrace the (momentary) support of an external principle of order and composition that, while affirming the notion of an uncompromised authority and the allegiance it commands and deserves, at the same time relieves him of both the responsibility and the freedom of standing alone.

But stand alone he must, until this position once again becomes unmanageable also, for the external authorizing principle will not prove any more enduringly adequate than the alternative of reliance on a personal vision. When the Trojan legend is finished, the narrator quickly praises the "noblesse" and "richesse" of what "I saugh graven" (471–73), but he is then left only with questions about the source of the story and about himself: "But not wot I whoo did hem wirche, / Ne where I am, ne in what contree" (474–75). To find some answers not available from his engraved source, he decides to leave the Temple, and the authority contained on its walls is for all purposes obliterated as he takes some assertive action and steps outside, only to discover that there is absolutely nothing to behold.[44] The old familiar phrase "then sawgh I" is carried out with him for support, but it refers now to a vacant landscape, devoid of any visible signposts:

> Then sawgh I but a large feld,
> As fer as that I myghte see,
> Withouten toun, or hous, or tree,
> Or bush, or grass, or eryd lond . . .
>
> (482–85)

The refrain, next modified by negation, points to the absence of anything other than the self, and to an external source that is not there:

Ne no maner creature
That ys yformed be Nature
Ne sawgh I, me to rede or wisse.
(489–91)

The sudden and complete lack of an organizing principle that can advise and guide provokes the narrator's anxious cry, " 'O Crist!' thoughte I, 'that art in blysse, / Fro fantome and illusion / Me save!' " (492–94). He resists this opportunity for creative independence and doubts in advance the legitimacy of anything he might produce out of his own resources to fill this blank scene. The Temple walls engraved with conventional literary "truths" could not furnish an adequate enough context for the narrator's voice, but the empty landscape unmarked by any prior or superior text inhibits and overwhelms the authority and autonomy of the individual mind that confronts it.

III

While in Book I, the text engraved on the Temple walls seems to vanish when the dreamer attempts to assume narrative control, the engraved images in Book III are literally disappearing: the melting names on the hill of ice, which forms the "feble fundament" (1132) for Fame's castle, offer an appropriately temporary testimony to an unstable history and tradition that supply only momentary supports. The description of the House of Fame forcibly recalls the narrator's activities in the Temple of Venus. The same phrase of reference—"tho sawgh I . . . ygrave," "ther saugh I" (see 1136, 1250, 1259, 1271, 1275, 1364, 1394, 1456, 1481, 1496, 1497)—is used consistently; here, however, we do not need a shifting narrative stance to indicate the inadequacy of the source, for the things described represent their own compromised status as reliable, legitimate models and guides. The catalogue of artists who adorn the interior and exterior of the palace—the "mynstralles / And gestiours, that tellen tales" (1197–98) and "hem that writen olde gestes" (1515)—portray the lack of any authoritative standard to negotiate among the different songs and stories they deliver. The minstrels are engaged in literary competitions (1127–32), and the poets debate about their various versions of the Troy story: since Homer is favorable to the Greeks, he is accused of "feynynge in hys poetries" (1478). The entire question of whether Homer "made lyes" (1477) in his poem suggests the same unsatisfied search for an absolute source of "trouthe" external to the individual author that was a primary concern in the Dido and Aeneas episode. After surveying the numerous literary representatives, the narrator responds to the confused mingling of their contradictory stories by throwing up his hands in a gesture of despair:

> What shulde y more telle of this?
> The halle was al ful, ywys,
> Of hem that writen olde gestes,
> As ben on trees rokes nestes;
> But hit a ful confus matere
> Were alle the gestes for to here,
> That they of write, or how they highte.
> (1513–19)

As if responding to the lack of any secure point of reference, the narrator turns his attention to the Goddess of Fame, who has begun to hold court. Set up as the reigning power, Fame quickly betrays her position as a reliable authority. Nine representative groups appeal to her, and although she issues firm decrees, they are completely contradictory and capricious—her declarations of good or bad fame bear no relation to the qualities of the people who must accept her judgment. But while Fame may be inconsistent in her response to the applicants, the applicants themselves are equally inconsistent in their requests: groups with similar qualifications ask alternately for good, bad, or no reputation at all. Fame's castle seems to be the destination of the dreamer's journey, but he finds here no adequate principle of order. The chaotic, arbitrary, and incomprehensible nature of what he views at *all* levels draws from him a statement of faith in himself and in his art:

> Sufficeth me, as I were ded,
> That no wight have my name in honde.
> I wot myself best how y stonde;
> For what I drye, or what I thynke,
> I wil myselven al hyt drynke,
> Certeyn, for the more part,
> As fer forth as I kan myn art.
>
> (1876–82)

This stands in the poem as the culmination of several less fully elaborated assertions of individual self-reliance that follow the debunking of conventional sources of authority. It is the final and firmest expression of faith in self as artist—as principle of conception, judgment, and interpretation—that the narrator delivers. As a statement of autonomy, this passage recalls and incorporates two earlier portions of the poem where similar issues are in question. One of the temporary claims of creative independence that we have discussed occurs in Book I at the moment of Aeneas's betrayal of Dido, when the narrator alleges "non other auctour" and then goes on to present Dido's complaint against the "godlyhede / In speche" that carries with it "never a del of trouthe." Faced with this condition, the narrator who has claimed full authorial independence creates a character who denies her own authority and command over the picture of herself that will be presented to the world. Dido admits that she lacks control and is, instead, subservient to the machinations of others:

> Now see I wel, and telle kan,
> We wrechched wymmen konne noon art;
> For certeyne, for the more part,
> Thus we be served everychone.
>
> (334–37)

The terms in which Dido disparages her understanding and mastery of "art" look forward to the later scene where the narrator, having witnessed some "godlyhede / In speche" but "never a del of trouthe" from the representatives of literary history who "tellen tales," as well as from Fame herself, responds quite differently. Rather than deny his autonomy and power, he asserts his own control over and through his "art": "Certeyn, for the more part, / As fer forth as I kan myn art."

The second passage echoed is from Book II, where a recognizable pattern of self-assertion and submission emerges. The pedantic eagle, an incessant talker who parodies himself, science, logic, rhetoric, as well as his literary precedents, overwhelms the mostly speechless narrator during the greater part of the journey through the air. But toward the end of the flight, when the eagle begins a new discourse on the "sterres" that will, he claims, aid his companion's comprehension "when thou redest poetrie" (1001), the narrator rediscovers his voice and is provoked finally to discount and repel these endless commentaries:

> "No fors," quod y, "hyt is no nede.
> I leve as wel, so God me spede,
> Hem that write of this matere . . ."
>
> (1011–13)

The narrator denies the utility of the eagle's recitations and in doing so discards the laws and rules that the eagle expounds. The recovery of his own voice—and his ability to control the discussion—coincides with his rejection of the learning that is both displayed and parodied by the eagle, but it is also accompanied by a substitution of another set of guidelines. It is, in fact, immediately followed by a reinstitution of the *auctores*—"hem that write of this matere"—as self-sufficient authorities that replace both the eagle's catalogue of lore and the momentary assertion of self portrayed in the narrator's decision to silence and resist the eagle.[45]

What immediately elicits this statement by the narrator is the eagle's suggestion that his passenger actually has no firm hold on the subject they are discussing:

> For though thou have hem ofte on honde,
> Yet nostow not wher that they stonde.
>
> (1009–10)

It is with similar words that the narrator, in Book III, claims the superiority of his own perception and judgment, his independence from external influence: "Sufficeth me, as I were ded / That no wight have my name in honde. / I wot myself best how y stonde."

The similarities of vocabulary and syntax serve to highlight the differences between the earlier statements that deny individual authority and this later passage that claims it—and that perhaps suggests the greater assurance developed by the narrator as he discovers the inadequacy of all external models. However, the similarities with the earlier expressions also function to connect them all into a unit where denial and assertion of autonomy become inextricably, almost reciprocally, merged. Despite the tone of confidence, the narrator's self-reliance in Book III is as fleeting as his pity for Dido and his condemnation of Aeneas in Book I, as temporary as his self-assertion in reaction to the eagle's accusation of inadequacy in Book II. Although the narrator's response to the activities of Fame has often been held up as a statement of complete detachment and of faith in self,[46] it is nonetheless a position from which the speaker retreats even as he voices it: after his initial claim, that "I wot myself best how y stonde," he admits that he is "certeyn" only "for the more part," and only "*as fer forth as I kan* myn art."[47] These qualifications, in combination with the echoes of earlier negative statements, place this stance in the middle ground between the forthright independence of "non other auctour alegge I" and the self-effacing subservience of "the book seyth." The confidence in the personal vision here is forceful but not total; the passage diminishes in impact as it continues, modifying itself by maneuvering into a careful balance between untenable alternatives. It is, however, a delicate balance, easily disturbed and, as the remainder of the poem indicates, not fully satisfactory. Fame's palace is unexpectedly denied status as the final destination of the journey, suggesting that the dreamer has not yet located what he was—and still is—looking for. His search for a reliable source and model for his art continues even after this statement, in which he asserts his superiority to and independence of external authorities while simultaneously creating (through his qualifications) a space that invites a more comprehensive and reliable standard than his own. The House of Rumor is one step further back toward an originating—and original—point; it is here that the narrator goes to locate the now confused and ill-defined object of his quest: "Somme newe tydynges for to lere, / Somme newe thinges, y not what . . ." (1886–87). This labyrinthine structure, where the true and false become fused and indistinguishable, qualifies even further the narrator's assertion of self-reliance and, for that matter, reliance on anything that professes knowledge or certainty. Furthermore, the development of "tydynges" as it is described in this section denies the possibility of true subservience, of transmitting without transforming with one's personal stamp. The final text, we discover, is constituted by the merger of what become inseparable contributions from individual authors:

> But al the wondermost was this:
> Whan oon had herd a thing, ywis,
> He com forth ryght to another wight,
> And gan him tellen anon-ryght
> The same that to him was told,
> Or hyt a forlong way was old,
> But gan somwhat for to eche
> To this tydynge in this speche
> More than hit ever was.
> ....
> Thus north and south
> Wente every tydyng fro mouth to mouth,
> And that encresing ever moo . . .
> (2059–67, 2075–77)

All that the narrator has endeavored to find and define meets its final dissolution in the House of Rumor. The Goddess Fame is, as Miskimin notes, "a parody of the images of authority that poets seek";[48] there are no criteria for—because there is no possibility of—distinguishing "fals" and "soth"; not only are reliable models missing, but the notion of truly copying and passing on a prior text without alteration is extinguished; and even the concept of the inviolate individual creation is similarly compromised. If the lack of a universal authoritative principle in the House of Rumor suggests that the individual vision must be self-justifying and self-sufficient, it also exposes the *need* for an outside standard that can provide an organizing principle for the unstructured, random mass of material that is found there. The absence of such criteria for order and validity does not, as some critics suggest, simply relieve the artist of responsibility or free him "from the obligation to make explicit choices in his art";[49] in fact, the declaration of independence of the personal artistic vision provokes a desire to escape the responsibility that is *conferred* by that freedom and autonomy. The dreamer is not a passive observer in the House of Rumor; he rushes among this fourteenth-century "miscellaneous rabble" and eagerly joins their frenzied acitivity ("I alther-fastest wente / About" [2131–32]), which is intensified by the sounding of a "gret noyse" (2141). All lesser, personal business is forgotten; all attention is now focused on the figure signaled by this sound, and all activity converges upon him. The response of the people indicates—and for some critics it suffices as confirmation[50]—that this man is the authority that has until now been withheld. But a somewhat wiser—or at least more cautious—narrator tells us that the new arrival, whom he "nevene [i.e., name] nat ne kan," is one who only "*semed for to be* / A man of gret auctorite" (2157–58; my emphasis). We know by now in the poem, even without the narrator's qualifying remark, that "authority" is, at best, a temporary quality; it might be helpful to remember that in a later poem Chaucer will introduce the duplicitous Calkas with a similar phrase—he is "a lord of gret auctorite"—only to transform him, in the space of three stanzas, into a "traitour."[51] The *House of Fame* breaks off, perhaps concludes,[52] before this ambiguous figure of authority speaks or acts, but the poem as it stands has worked to show that no authority can maintain its status or offer an unassailable system upon which an author may safely rest when called upon to justify the legitimacy of his art.[53] There is no reliable source of guidance in the poem: neither an external universal figure or model, nor the narrator himself, who is constantly left disappointed and unsatisfied by all potential props for his vision. As each is found inadequate, he begins tentatively to become aware of his own position—to test, rely on, and assert the power of his own perspective. But as soon as these impulses emerge, new conditions arise that qualify his status and that suggest the need for a larger, more comprehensive, and stronger—because less individual—source of sanction.

The complex implications of the narrator's final situation at the conclusion of the *House of Fame* can be further elucidated by reference to another of Chaucer's dream visions, the *Parliament of Fowls*. In this poem, a (presumable) voice of "gret auctorite" is introduced earlier and as a more central figure: the Goddess Nature presides over the parliament of fowls who have gathered to choose their mates on Saint Valentine's day in accordance with Nature's ordinance. But when Nature fails to provide an authoritative judgment concerning the crucial choice of the formel's mate, and instead transfers the responsibility of rendering a "verdict" onto the council of birds (522–25)—and later onto the formel herself (620–22)—she reveals the inadequacy of her own governing authority.[54] It then quickly becomes apparent that this ruling goddess can neither effectively guide the debate generated by the competing voices of different factions among the birds, nor, when she simply interrupts the debate, sufficiently guide or determine the individual choice of the formel, who then refuses to resolve the issue by providing a judgment of her own. At one point during the discussion opened up by Nature at the beginning of the parliament, the "kokkow" decides to take charge and solve the problem, citing as his credentials "myn owene autorite" (506). The implicit acknowledgment of the limitations of Nature's authority is thus met with the assertion of authority from an individual voice. But this self-assertion is immediately met by the "turtel's" rebuttal that "bet is that a wyghtes tonge reste / Than entermeten hym of such doinge, / Of which he neyther rede can ne synge" (514–16).[55] The recommendation is to be silent rather than to act on the shaky grounds of individual authority, to say nothing rather than to encounter the inevitable limitations of one's personal voice.

This response counsels a silence of defeat—the cessation of speech when there is nothing legitimate to authorize it. Yet, to step back from the poems for a moment, this silence of defeat finds its reverse image in the Augustinian notion of a more triumphant silence: the "rhetoric of silence,"[56] the internal silent words through which "truth" is expressed and delivered, toward which all language tends and in which it culminates. Language, the spoken word, the expression of a personal voice, the "words instituted and used by men"[57]—this language is to be subsumed finally in the movement from human voices to the silent and fully authoritative voice of the true creation and creator. The silence counseled to the "kukkow" in the *Parliament of Fowls* is not of this variety, but in the *House of Fame* there is a wordless conclusion—when the narrative voice abruptly ceases at the appearance of a man who only seems to be a figure of authority—that embraces and negotiates both of these silences. Throughout the poem the narrator has alternately asserted his superiority to and attempted to disappear behind previously authorized versions of "trouthe," claiming and disclaiming his role as author as he tests and rejects all available authorities, including himself. No position—whether of self-effacing dependence, or assertive self-reliance, or even a more moderate middle position—has proved satisfactory. What results is an oscillation among inadequate options that constantly surface, call each other into question, and replace each other. The narrator's silence at the end acknowledges the limitations of his own voice but also recognizes and predicts the inevitable failure of the "man of gret auctorite" to maintain his status, and it recalls the evidence of similar failures that have occurred throughout the poem. No truly authoritative voice has been discovered; no authorial stance can be fully validated; and the poetic voice and enterprise end. This silence, however, is not entirely hopeless. Because it occurs just as the man of great authority appears, it keeps before us this figure's *potential* power to carry the full weight of the authorial position—to do it so completely, in fact, that he leaves the narrator no space in which to function. The silence, then, also serves as a gesture of deference and allegiance to this authority—or, to be more exact, the silence serves as a gesture that *confers* authority by refraining from subjecting the "man of gret auctorite" to scrutiny and to the challenge of a competing voice. If the penultimate moments of the poem expose the need for a reliable external source of authority, the narrator responds to that need in the final moment. The Augustinian notion of the subsumption of the individual voice by the truly authoritative one takes on a different dimension in this deliberate act, which circumscribes and compromises personal authorship in order to salvage the image of an anonymous figure whose authority is merely suggested, and never tested.

CHAPTER III

Unaffirmed Art: Allegory's "Unjust Possession"

I

In Skelton's *Bowge of Courte,* a fifteenth-century dream vision, the act of writing is openly announced as a major and motivating concern. Comparing himself to the "greate auctoryte / Of poetes olde," the speaker is "sore moved" to try his hand in "this arte," but is "not sure" whether he should—and can—write.[1] Struggling with this dilemma, torn by indecision ("Thus up and down my mynde was drawen and cast / That I ne wyste what to do was beste"), he falls asleep and dreams that he is on a ship, an allegorical image of the court. In his vision he encounters a series of "full subtyll persones" who purport to offer him guidance and authoritative advice, but all of them are revealed as duplicitous, corrupt, and untrustworthy. During the dream, the speaker is systematically betrayed by the entire cast of characters, who plan, at the end, to "slee me of mortall entente": these false figures, creating an environment of widespread uncertainty and arbitrariness, threaten to annihilate him completely. As they descend upon him, the speaker (who has not yet claimed for himself the title of poet) seeks to escape from them by jumping off the boat (and, since the boat is the dream setting, by jumping out of his dream as well)—symbolizing his rejection of these figures and his attempt to detach himself from them. This desperate but deliberate act of self-preservation is also potentially an act of self-destruction, which is, in the end, transformed into the act of writing:

> the shypborde faste I hente,
> And thoughte to lepe; and even with that woke,
> Caughte penne and ynke, and wrote this lytyll boke.
> (ll. 530–32)

The search for authoritative guides may be as unsuccessful in this poem as it is in the *House of Fame* (as Stanley Fish writes, "In the end the interpretation and the value of this dream remain problematical because it offers neither the reader nor Dread anything authoritative"[2]), yet in the *Bowge of Courte* this need to erect but failure to locate an authority does not culminate in a self-effacing silence; instead, it is what enables and motivates the poet to write. The problem posed at the beginning of the poem—whether (and how) the speaker should assume the "greate auctoryte / Of poetes olde" and try to write—inspires the dream, which provides no direct, concrete answer but nonetheless produces the uncontested impulse (and the apparently uninhibited ability) to write "this lytyll boke." The details of the original question are abandoned, while the basic problem is almost magically resolved: a defiant yet desperate and dangerous decision metamorphoses the uncertain speaker into a confident poet unperturbed by the debilitating indecisiveness that had initially plagued him. The *Bowge of Courte* ends with a poet who neither lapses into silence nor, still uncertain, continues to seek but who finishes and offers to his readers a completed poem that has as its subject and structure the unfulfilled quest for an authoritative voice.[3] The problem has become the solution, and the speaker has become a writer: the absence of authority has become the material of the poem, and it has made the poet.

The idea that the failure to locate an authorizing principle can produce such an advantageous situation for the poet finds a more extreme formulation in Sidney's *Apology*. When Sidney answers the charge that poets are liars, he contends (following Boccaccio, and followed by Harington)[4] that "to lye is to affirme that to be true whiche is false."[5] Other artists (e.g., historians, astronomers, geometricians, physicians) can "hardly escape from many lyes," for they "take upon them[selves] to affirme." In other words, only those who assert their words as true, who present them as confirmed and ratified, can be liars. But the poet, unlike these other artists, "never maketh any circles about your imagination, to conjure you to beleeve for true what he writes," and thus, Sidney concludes, "for the Poet, he nothing affirmes, and therefore never lyeth" (pp. 184–85). This argument attributes to the poet a peculiar position of authority. It is precisely the absence of authorization that gives the poet his privileges and, in fact, that sanctions what he writes; he achieves his authoritative status (he can never be accused of lying) because he never authorizes anything.

This answer to the charge of lying seems to provide easy absolution for the poet, assuming that he would be able to create a fiction that could not be confused with fact, that was clearly not affirmed. According to Sidney, this feat too was simple, for the distinction was an obvious one—one that even a child could recognize: "What childe is there that, comming to a Play, and seeing *Thebes* written in great Letters upon an olde doore, doth beleeve that it is *Thebes*?" (p. 185). Thus, although one reads a history "looking for trueth," one should read "Poesie, looking for fiction" (p. 185), since, unlike the "meaner sort of Painters (who counterfet onely such faces as are sette before them)," the "more excellent" poet has "no law but wit'': he borrows ''nothing of what is, hath been, or shall be,'' and makes no pretensions that he does (p. 159). He ''recount[s] things not true'' and—this is what exonerates him—''he telleth them not for true'' (p. 185).

But this raises several questions about the poet's function. In his analysis of the debate between fact and fiction in the Renaissance, William Nelson writes, alluding to Sidney, that the poet could free himself from the charge of lying by ''overt or tacit admission that the story was indeed fiction and therefore not subject to judgment as to whether it was historically true or false.''[6] One way of making this admission, adds Nelson, was ''by the obvious difference of the tale from history'': in other words, by making clear that the fiction was not presented as an image of the actual world, that it was a picture totally distinct from the picture human experience presents to us, so that there could be no mistaking the fiction for, to use Sidney's phrase, an ''actuall truth'' (p. 185). But as Nelson notes, ''admittedly invented story had to find its raison d'être on grounds other than those which gave true history its value.''[7] In short, the poet who removes himself from the charge of lying by insisting upon his text as fiction becomes vulnerable to the ''charge of frivolity'':[8] if he cannot claim fidelity to ''what is,'' he must find some other justification to legitimize his art. This justification was, for Sidney and others, a moral one. Poets present not the ''brasen'' world of man's ''infected will'' but the golden world of the ''erected wit'' that ''maketh us know what perfection is'' (pp. 156–57); they do not replicate what men see and do, but far surpass this and ''bestow that in cullours upon you which is fittest for the eye to see'' (p. 159): not ''what is, or is not, but what should or should not be'' (p. 185). What saves acknowledged fiction, then, from the charge of frivolity is its moral function: exempt from fidelity to the truth of experience, the poet can create a moral truth not available from actual history.

(Since man and the world are seen as fallen, fiction and moral truths come to occupy the same space.) Thus both issues seem to be resolved simultaneously: the poet, by intentionally and overtly departing from the conditions and contingencies of the actual, produces a perfected world that is unmistakably invented, that does not—and by definition cannot—correspond to fact.[9] Fiction is therefore *sanctioned* by its immediately recognizable contrast to fact: indeed, the efficacy of the poet—whose function lies in his ability to invent a morally ideal landscape not found in the world of human activities, not normally accessible to the infected will—depends upon the discrepancies readily discernible between his text and "what is, hath been, or shall be."

But this, too, raises problems. First of all, the idea that fiction must define itself in contrast to the world suggests that it *needs* the context of that world to identify its own domain; in other words, the notion of separateness implies inseparability. Furthermore, and even more crucially, if there is such an obvious disjunction between the virtuous model of what should be and the actuality of what is, then the poet's world is liable to become withdrawn and isolated, so far removed from the behavior of men that it is purely visionary, with no real power (to use Sidney's phrase) to "move men to take that goodnes in hande" (p. 159)—since the goodness is too distant to be grasped. When the unaffirmed is in this way presented as forever unaffirmable because so completely detached from human experience, it denies what Sidney calls "the ending end of all earthly learning"—that is, "well dooing" or "vertuous action" (p. 161).[10] Sidney avoids these consequences by coupling his stricture of "what should be" with "what may be" (p. 159)—his addition to the Aristotelian requirement[11]—thereby mandating that the ideal be presented as potentially realizable.[12] Though this criterion brings poetry more in line with its moral function, however, it breaks down the barrier erected between what the poet eschews (what is, has been, or will be) and what he invents (what could and should be), for if man can or might be moved to act morally in accordance with the ideal vision, then "what may be" is what possibly "shall be." And once these categories are brought closer together, then the distinctions that they formed between unaffirmed fictions and experiential reality become blurred. When the poet's picture of what should be is joined to what could be, there must be some correspondence between his fiction and man's world; and when this correspondence occurs, it is no longer possible to detach completely the fiction from that world, to claim the unaffirmed nature of that fiction, to declare that "things not true" are recounted and that they are recounted "not for true"—in short, it is no longer possible to create so great a discrepancy that the fiction cannot be confused with fact.

This crux is further exacerbated by Sidney's acknowledgment of the charge that poets employ a style that "argueth a conceite of an actuall truth" because they "gyve names to men they write of." But this, Sidney argues, is not done "to builde any historie": rather, poets must do it "to make theyr picture the more lively" (p. 185). They invent the pattern of exemplary figures but must then give them substance, granting them a local habitation and a name. Sidney indirectly admits here that although the poet does not plan to present his fiction as actual, he must rely upon the discourse and structure of the actual to convey his ideas; he must, Sidney implies, conform to the human mode of understanding and employ its language. The poet's fictions, then, no matter how different from the brazen world they are intended to be, are bound by the form of that world and must be expressed in a form very similar to it. Again we find that the autonomy of the poet and of his fictions are being circumscribed. If the idea that fictions must be defined in relation to the world suggests dependence, this recognition of their moral function and linguistic limitations further compromises the notion of detachment and difference so crucial to their defense.

The authority acquired by lack of authorization is, thus, an unstable and dubious quality that lends an equally unstable and dubious legitimacy to any poetic text. Even Sidney seems to acknowledge the difficulty of keeping the poet within the limits of the unaffirmed, and near the end of the *Apology*, he appears to subvert his initial judgment when he admits that poets make a crucial mistake: "where we should exercise to know, wee exercise as having knowne: and so is oure braine delivered of much matter which never was begotten by knowledge"; the poet's "*Quodlibet*" "matter" is rendered useless, since, "never marshalling it into an assured rancke," the poet loses rather than instructs his readers, who "almost . . . cannot tell where to finde themselves" (pp. 195–96). The emphasis on knowing as opposed to "as having knowne" implies an estimation of what is valid and confirmed over what is merely assumed or asserted to be so, and while this judgment reinforces the caution that poets should not "tell for true" their fictions, it also replaces that requirement with a higher one. Now a more objectively oriented knowledge becomes the matter of the true poet,[13] and this belies the former edict that not only exempted the poet from the verifiable world but also made mandatory that he refrain from engaging in it. Authority achieved for the price of authorization is, indeed, an ambiguous distinction. It prevents the poet from presenting his text as anything but a fiction, yet while it may not "conjure you to beleeve for true what he writes," it does, for that very reason, conjure up the known (or at least knowable) world that lies outside of that unlying but unaffirmed vision—a known world that must, then, be incorporated into what can no longer be labeled as pure fiction, thereby destroying that protective domain of the unaffirmed that could supply the poet and his text with authority.

The attempt to validate poetry as fiction by removing it from actual experience concludes by pushing it closer to that experience; the authority of the invented and unaffirmed is replaced by the substantiality of the real, the actualizable, and the verifiable. This curious evolution seems to have special significance for a particular mode of expression, for one feature endemic to most Renaissance responses to the charge of lying is that they inevitably include or lead into a discussion of allegory. For Boccaccio, Sidney, and Harington, for instance, allegory is the prime example of poetry that "telleth not for true" and therefore does not lie. Allegory is, in fact, presented as the mode of the unaffirmed: according to Sidney, "they [the poet's "persons and dooings"] will never give the lye to things not affirmatively but allegorically and figurativelie written" (p. 185)—and thus do not give the lie at all in any true sense. Allegory is consequently plagued, however, by all the contingencies and complications that qualify the efficacy, authority, and even the existence of unaffirmed art, which must be fictive and distinct from the verifiable world but whose ethical purpose makes this separation self-defeating, even impossible. Chapman, following his predecessor, ties all these elements together in a single passage:

> Nor is this all-comprising Poesie phantastique, or mere fictive, but the most material and doctrinall illations of Truth, both for all manly information of Manners in the yong, all prescription of Justice, and even Christian pietie, in the most grave and high-governd. . . . if the Bodie . . . seemes fictive and beyond Possibilitie to bring into Act, the sence then and Allegorie (which is the Soule) is to be sought—which intends a more eminent expressure of Vertue, for her loveliness, and of Vice, for her uglinesse,

> in their severall effects, going beyond the life than any Art within life can possibly delineate.[14]

In other words, allegory is not "within life," and its freedom from that constraint enables it to fulfill its moral function, but allegory thus only "*seemes* fictive"; its moral function belies this appearance for, in fact, it is not "beyond Possibilitie to bring into Act." The art of allegory, then, as unaffirmed art that never authorizes, must be outside of life; but it must also be very much within life, for it ultimately locates its authority in an actuality that can confirm it.

This contradictory movement first away from and then back toward the reader's world of "what is" also informs the conventional definitions of allegory as a form that uses a fiction to veil an inner truth. Boccaccio, who provides the standard line, explains that the poet's job is to conceal, even "protect as well as he can and remove them [the truths] from the gaze of the irreverent, that they cheapen not by too common familiarity."[15] This argument forms the basis of a remarkably antireformative, exclusive theory of poetry: the hidden sense will be corrupted if it becomes the property of too many readers, and thus to preserve its purity, the veil is constructed so that the true meaning can be comprehended by—but only by—a small and elite group. Only the select few will be edified; the others will be content with the surface story. Thus, the allegorical poet's endeavor succeeds only when it fails on a grand scale: that is, it succeeds by preventing most of its readers from understanding it.[16] Although this theory locates a moral content behind the poetic fiction, it ultimately denies any truly moral *function* to allegory, for its moral intent is rendered inaccessible to the majority of men, who are—and therefore remain—coarse, profane, and debased. Inherent in allegory, then, is an acute awareness of the unredeemed and unredeemable nature of most of the world—a world that it tries to fend off, a world that it does not even try to reach, a world from which it tries to protect itself.

But there is more to Boccaccio's explanation, for within his definition of allegory he presents his defense of it, one that attempts to soften the rigid barriers erected between the allegory and its audience. As if anticipating objections to this privileged allegorical domain, which admits only a few, already enlightened men, Boccaccio claims that "Surely no one can believe that poets invidiously veil the truth with fiction, either to deprive the reader of the hidden sense, or to appear the more clever"; instead, he now states that they hide the truth only to increase its value, to reveal it in a way that will make it precious: "to make truths which would otherwise cheapen by exposure the object of strong intellectual effort."[17] The defense that poets do *not* try to deprive their readers of the truth blatantly opposes the idea that allegory removes and keeps the truth from the debasing touch of the coarse multitude of readers.[18] It suggests the writer's uneasiness about maintaining distance between his text and its audience, and it turns allegory into a mode that not only should reach its readers but also should require (and perhaps teach) them to exercise their minds in ways that they are capable of, if not accustomed to. In short, allegory, as it is conventionally defined, grounds its defense in terms that contradict its premises, and justifies itself by confirming its accessibility to the world it initially detaches itself from.

In the hands of the sixteenth-century rhetoricians, definitions of allegory exhibited a similar quality: initial requirements that the form and content of allegory be divorced from the world dissolve into requirements that they move closer to the world. The handbooks generally subscribe to the notion that in allegory a veil covers the inner sense, though in different terms: they claim that allegory says one thing on the surface but that another meaning lies behind the surface. Peacham's description is representative: "Aligory is a sentence, which sheweth one thing in wordes, and another in sence."[19] But Peacham's example of allegory from Cicero presents a notion of that covering veil that blurs the distinctions between the levels, for it is one "whose signifycation is commonlye knowne"; in other words, he stresses the easily transparent, not the potentially impenetrable, nature of the literal level.[20] This attitude, found in most of the handbooks, raises the question of whether allegory's primary function is to keep remote a truth that would be sullied by the common understanding, or to make a remote truth more accessible to the common understanding. When the rhetoricians—as they all do—further define allegory as a continued, extended, or perpetual metaphor, they turn it into a mode in which there is a great affinity between the sense and the literal, between the fiction and the underlying truths. For the definition of metaphor almost always stresses similarity: according to Peacham, for example, metaphor is "when a word is translated from the proper and natural signification, to another not proper, but yet nie and likely." Thus, while it is agreed that the surface depicts one thing and the hidden sense another, the relationship is more one of similitude than of contrast, of resemblance than of difference. And the reasons given for using metaphor enforce this idea that allegory ultimately reveals rather than conceals. It is employed sometimes for "pleasauntnesse," but also for "necessity": in other words, often no word can accurately and directly express the meaning, and so poets are forced, in order to make their meanings clear, to employ these figures that do not exactly fit but that supply the closest fit possible.

Puttenham's analysis provides the most interesting case, for it exposes all the various movements away from and toward the notion of allegory as an exclusionary, deflecting, and duplicitous device, and it concomitantly reveals much of the ambivalence that informed the Renaissance theory of allegory. According to Puttenham, allegory is employed "when we speake one thing and thinke another, and that our wordes and our meanings meete not."[21] This defines it as essentially duplicitous, as Puttenham makes explicit when he names it after this characteristic: "this figure therefore which for his duplicitie we call the figure of [*false semblant or dissimulation*]." But this straightforward acknowledgment of the dissembling nature of allegory is soon retracted. Although he initially locates dissimulation in "wordes [that] beare contrary countenaunce to th'intent," he retreats from this definition and contends,

> But properly and in his principall vertue *Allegoria* is when we do speake in sence translative and wrested from the owne signification, nevertheless applied to another not altogether contrary, but having much conveniencie with it as before we said of the metaphore.
>
> (III.18; p. 197)

Apparently dissatisfied with his opening definition, which undeniably specifies a misleading outward appearance, Puttenham here corrects himself to say that the words do not bear a meaning wholly different from the sense; in fact, his allusion to his own definition of metaphor invokes his emphasis, in that section, on the relation between the word and the signification being one "of some affinitie or conveniencie" (III.17; p. 189). For never is a metaphor employed to deceive or dissemble. The three possible reasons for using this figure are, as listed by Puttenham: (1) "for necessitie or want of a better word," and "for want of an apter and more naturall word," so that it "doth not so much swerve from the true sence, but that every man can easilie conceive the meaning thereof": i.e., when there is no better way to express something so that all men can understand it; (2) "for pleasure and ornament," though in this case the metaphors still "approch so neere and so conveniently, as the speech is thereby made more commendable"; and (3) "to enforce a sense and make the word more significative" (III.17; pp. 189–90). Thus, all three functions of metaphor stress that it is used to express meaning in the most clear and forceful way possible and to make that meaning clear and forceful to "every man." While in his formal definition Puttenham states that in dissimulation "the words beare contrary countenaunce to th'intent," his ultimate impulse is to reverse this and to define the figure called "Dissimulation" as one in which the words bear very similar countenance to the intent—one that does not so much conceal the meaning "under covert and darke termes" as it reveals the meaning through words that have some easily discernible "affinitie" with it, that are used when no "apter" words are available.

Allegory, then, expresses things for which we have no adequate terms; it has no access to a special language and must use words wrested from their own signification to convey the inexpressible ideas that can only be approximated verbally.[22] But since allegory is bound to the terms of conventional discourse, the two can never be clearly distinguished or divorced, and this negates any privileged sanction allegory would gain as an independent domain. Dissembling, reports Puttenham, is speaking "otherwise then we thinke, in earnest aswell as in sport, under covert and darke termes, and in learned and apparant speaches . . . and finally aswell when we lye as when we tell truth" (III.18; p. 197). As Isabel MacCaffrey notes in explicating this passage, "Allegory is a double-edged weapon, serving both lies and truth."[23] Because allegory is bound by the limits of human language, none of the distinctions that Sidney tries to erect in the *Apology* can survive; the allegory that in Sidney's initial definition could never lie because it never "telleth for true" here becomes, as it finally did in the *Apology*, a mode that can lie as well as it can tell the truth.

My purpose in grouping together Sidney's free-ranging poet, Boccaccio's divinely inspired poet, and Puttenham's handbook-bound poet-orator has been to highlight the characteristics shared by Sidney's claim for unaffirmed fictions and some standard and contemporary concepts of allegory. In all cases the attempt to justify the poet's allegorical mode initially takes the form of openly acknowledging and accepting its distance from the world of its readers, but eventually the justification is sought *in*—or in allegory's accessibility to—that world.[24] Setting it in opposition to things written affirmatively, things "told for true," Sidney first defines allegory as an autonomous world of fictional discourse that is approved because it *is* fictional and autonomous; it by definition seeks no authorization from the verifiable, fallen world of "what is"—in fact, it is authoritative precisely because it detaches itself from the "what is." Yet the substantive, confirmable nature of that world exerts its pull and control: the authority of unauthorized fictions is not complete, for the unaffirmed, the fictional, must be affirmable, actualizable.[25] To avoid becoming what one critic calls "the frenzied and solitary builder of castles in the air,"[26] the poet must work, to use Sidney's term, "substantially" (p. 157). And the more the poet is impelled to make his golden world comprehensible and realizable, the more he must accommodate the facts of human experience and human modes of perception and expression.[27] Thus the attempt to ensure that the fictions do not become remote requires that they be expressed through the language and structure of the actual, but since the actual is so far from the ideals of the fiction, this requirement makes it impossible to present the ideal fictions accurately in any definitive form, and hence, in a very different way, guarantees that they remain remote, even from the poet.

Allegory is born out of the recognition that man's world and his language are fallen.[28] Consequently, the fictional, the dark conceit, the words wrested from their natural signification, are upheld as the necessary means by which the writer can move beyond the confines of that fallen idiom and by which he signifies that he has done so. But his use of these means also suggests how tied to that world the poet is; how much he must work through, not apart from, that world; how much, therefore, he can only approximate, rather than fully embody, other worlds. The fiction and the dark conceits, then, become indications not of his freedom from, but of his involvement in and restriction by, the contingencies of the actual.

By the very methods used in the attempt to present and authorize allegory as distinct from "what is," allegory thus becomes defined as a part of "what is." Its criterion becomes not its distance from the actual, but its proximity to it; not what it can invent outside of that world, but what can be verified within it. All this works to destroy the autonomous domain of unaffirmed fiction and, therefore, its self-sufficient authority, ultimately denying the poet the privilege and possibility of authoring his invented ideal worlds. Allegory does not so much free the poet from the world as bind him to it, demonstrating that he and his art are confined and fashioned by it.[29]

II

In the *Mutability Cantos* of *The Faerie Queene*, one of Spenser's first decisions is to delay his examination of "How *MUTABILITY* in them [all mortal things] doth play / Her cruell sports, to many mens decay"[30] and, instead, to present a historical survey of some basic background information:

> But first, here falleth fittest to unfold
> Her antique race and linage ancient . . .
> (VII.vi.2.1–2)

These are established facts, we are told, that have already been recorded, and the narrator promises to follow his source faithfully:

> As I have found it registred of old,
> In *Faery* Land mongst records permanent.
> (VII.vi.2.3–4)

Although the narrator has already acknowledged the omnipresence of *"Change,"* which causes dissolution and decay, he here ascribes permanence to the historical records. The national archives of Faeryland, it seems, are unique in their stability, and the narrator identifies his narrative with them. At the outset of this final section then, the narrator, associating Faeryland with the permanent records and old registers of history, citing fictional history as his authority, dissolves the boundaries between fiction and fact.

To appreciate fully the significance and implications of this, we must look back, for this opening reference to historical records in the *Mutability Cantos* recalls and demands consideration of the other "antique Registers" mentioned in *The Faerie Queene*: in particular, the *"Briton moniments"* read by Arthur and the *"Antiquitie* of *Faerie* lond" read by Guyon in Book II, after their tour through the House of Alma. It is here, with the juxtaposition of these two stories, that we find Spenser's most sustained early attempt to define and defend the status of his Faeryland and, in the process, to delineate the provinces of fact and fiction, history and poetry, the actual world and his allegorical world. Yet it is also here that he establishes the conditions of their eventual disintegration and, therefore, the disintegration of Faeryland as well.[31]

The discrepancies between the British history and the account of Faeryland's past have been duly noted by critics: the *Briton Moniments,* sixty-five stanzas long, is a history full of violence, betrayal, upheavals, and interruptions in the line of succession; the *Antiquitie of Faerie lond,* only seven stanzas long, presents a peaceful, harmonious history with an unbroken line of succession to the throne. The differences between the two histories have been likened to the differences between the real and the ideal, history and poetry, fact and fiction, life and literature; the two chronicles have been called "contradictory" and "utterly irreconcilable."[32] Yet a crucial aspect of the histories is ignored when critics set about giving a general description to each. For although the British history is full of disorder and violence, it begins relatively quietly: the land is "all desolate," unsought even by merchants out for profit (II.x.5.6–8). Granted, it is a "salvage wildernesse" (x.5.3), occupied, we soon learn, by a "salvage nation" of "hideous Giants" who "Polluted this same gentle soyle" (x.7.1–2; x.9.2), but the real history of Britain does not begin until the arrival of the first British ruler, Brutus—a man who gains his rightful title through orderly inheritance ("anciently deriv'd / From royall stocke of old *Assaracs* line" [x.9.6-7]) and who rightfully and justly reclaims the land from that hideous, plundering race: "And them of their unjust possession depriv'd" (x.9.9). British history commences, then, with a lawful act of repossession by a man who lawfully derives his existence, title, and power from the line of succession: only afterward do the incidents of upheaval and rebellion occur.

On the other hand, the history of Faeryland, however peacefully it proceeds, begins disruptively with an act of "unjust possession." The first thing Guyon reads about is the creation of the Faeries themselves:

> It told, how first *Prometheus* did create
> A man, of many partes from beasts derived,
> And then stole fire from heaven, to animate
> His worke, for which he was by *Jove* deprived
> Of life him selfe, and hart-strings of an Aegle rived.
> (II.x.70.5–9)

The Promethean legend is here presented to account for the creation of "all Elfin kind," a race whose very existence derives from a stolen gift that (to borrow a phrase from Denis Donoghue's study of the "Promethean imagination") has its "origin in violence, risk, and guilt" and that implicates its recipients in the original crime of theft.[33] Prometheus's action, defiant and blasphemous, animates the Elfin clan, giving them life, but a life that is inherently contentious, usurped, and unauthorized, an existence reaped as the spoils from an act of theft: only afterward does the orderly progression of events unfold.

If, then, Faeryland is the world of fiction, of imagination, then fiction and the imagination exist only in opposition to a higher order and have at their source a rebellious, defiant quality. In fact, these qualities *define* the act of creation that can produce the triumphant vision of the truly virtuous, orderly life—a vision that is (like the Promethean fire) an unauthorized possession. Creation itself and even the noble fruits of creation derive their existence, substance, and force from an acknowledged absence of authorization.

This Faeryland—a created and essentially unauthorized realm—also seems defined by its difference from the world of human experience about which Arthur reads, a world that it is clearly most unlike. I do not deny the claims of other critics that the Faeryland chronicle is quite distinct from the British one. In fact, I think the obvious differences are courted and made explicit by the poet (it is not insignificant that Guyon and Arthur each read alone, neither of them seeing what the other one so avidly peruses, neither of them sharing with the other the substance of his text). By so insisting that the stories are separable, the poet guarantees himself and his Faeryland a space to inhabit where there are no pretensions to (and hence no demands for) correspondence to the structure of actual experience—where, therefore, he can freely invent his ideal moral landscape without making it or himself vulnerable to charges of lying, of infidelity to what is. His invented world is admittedly unaffirmed, obviously not fact, clearly unverifiable by the standard criteria of authenticity; and this is made clear by the discrepancies between it and the account that precedes it.

But it is with the overreductive distinctions made by those critics that I disagree, because there is a more complex side to them that even my description above obscures. The relationship between human experience and the fiction *is* one of explicit contrast, but it cannot be characterized by general harmonious/disruptive, peaceful/violent categories subsumed by a simple ideal/real dichotomy. Because of the dual nature of both stories (their curious beginnings that do not fit the general content), the contrasts suggest dynamic alternation, not static opposition. The two worlds are, in fact, mirror images of each other: the first with an orderly beginning, followed by disruption, rebellion, and usurpation; the second starting with an act of defiance and usurpation, then proceeding peacefully, without disruption. Their exact contrasts imply that they are reciprocal states, and together they signify a continuum on which they both exist: harmony dissolves into disharmony, disharmony evolves into harmony; the ideal and the real can slip into each other. Therefore, although they now occupy different positions on this scale, we cannot conclude that these two histories are totally disjoint, that they are "utterly irreconcilable" views that can only be juxtaposed but never integrated.

On the one hand, the similarities and the dynamics evoked by the twofold nature of the contrasts indicate that the ideal state is potentially realizable, that it is an actual possibility. On the other hand (and more disturbingly), they present the Faeryland vision as ultimately unattainable, predicting that it will recede, become remote, disintegrate. For this continuum suggests what is never directly expressed here: that the ideal vision of peace is not eternal; it is not as disassociable from reality as it may first appear, and thus it, too, may eventually dissolve into discord. In fact, the brevity of the account itself, compared to the lengthy British chronicle it follows, may imply that its vision cannot be sustained for very long. It is as if the origin of that stolen gift of Faeryland life never relinquishes its hold; the force of that defiant, unauthorized act always exerts itself and will continually be felt, will always be a part of the harmony that it both provides and defies. And the harmony will always contain elements of the discordant and troublesome reality that it inverts and needs as a foil. The only true act of creation described in both histories is the one that begets the Elfin race, an act of creation that is inherently usurped and disruptive but that also allows the ideal to be formed *out* of the disturbing reality—an ideal, however, that is short-lived, that is born from disruption and carries it within itself.

If there is, then, a rift between the two histories, it is a rift between a fictional creation (obtained at great cost and risk) of an idealized version of human history and a more accurate description of that history—an ideal not simply set against the world, but an ideal *of* that world, sharing several of its attributes and presenting, briefly, a complementary picture of it.[34] They are, in short, two sides of the same coin. The disturbing elements of this relationship are kept in the background here: Spenser concentrates on the positive—Faeryland does figure forth the ideal—and he emphasizes the beatific product of creation rather than its less sanctified source. He capitalizes on the positive function of the contrasts that allow Faeryland to be constructed and on the positive value of the reciprocity those contrasts imply: in Faeryland, it is possible to embrace the ideal order, to capture it and hold it up as a version of the world of human endeavors. But although this passage is, for the most part, optimistic about the poet's (as well as man's) access to that vision, its less attractive qualities—the negative implications of Faeryland's dependence on the human world as a counterpoint, of the dynamic relationship this dependence necessitates, of the characteristics Faeryland must then share with actual human existence, and of the potentially debilitating nature of the act that generates it—are unmistakably and inescapably present.

The darker side of this concept—the difficult, unstable, disruptive nature of the Faeryland harmony—moves more into the foreground in a later passage that likewise sets together two versions of a similar story. At the beginning of Canto xi of Book IV, the narrator returns, after a lengthy absence, to the story of Florimell and Marinell; uncharacteristically, he comments on his own activity, berating himself for (characteristically) leaving parts of his story unfinished.[35] He has "Left a fayre Ladie languishing in payne"; he has "doen such wrong, / To let faire *Florimell* in bands remayne" (IV.xi.1.2–4). But this self-castigation is misdirected, for the narrator next tells us that the crucial question, for which even he has no answer, is whether the story can or will be completed: it is out of his hands now, he adds, and only an unpredictable power more mighty than his as author can redeem the situation. After the grand introduction of promised resolutions, he can only lamely express his pity for Florimell:

> From which unlesse some heavenly powre her free
> By miracle, not yet appearing playne,

> She lenger yet is like captiv'd to bee:
> That even to thinke thereof, it inly pitties mee.
>
> (IV.xi.1.6–9)

The situation, in fact, seems rather dismal and hopeless. Florimell is held captive by "Unlovely *Proteus*," who plans "thereby her to his bent to draw"; she is imprisoned in a "dongeon deepe and blind," "cruelly" chained, dwelling with "horror" (xi.2–4), and, even worse, she continues to refuse Proteus out of unrequited love for someone who himself shows no signs of relenting: "all this was for love of *Marinell*, / Who her despysd" (xi.5.1–2).

Although the narrator has begun this canto with the stated intention of returning to redeem Florimell after a period of unconscionable neglect and delay, all he can do, at first, is describe her situation, offer his sympathy, and then leave her once more. This time he abandons her not to pursue the story of other Faeryland characters but to visit a more naturalistic and mythological setting and to describe the wedding of the Thames and the Medway. A bit of the history of their courtship is provided first: the Thames had tried to woo the Medway "to his bed," "But the proud Nymph would for no worldly meed, / Nor no entreatie to his love be led; / Till now at last relenting, she to him was wed" (xi.8.6-9). Initially, two parallels to the story of Florimell are set up here: Proteus's as-yet-unsuccessful attempt to win Florimell, and Florimell's as-yet-unfulfilled relationship with Marinell. But the details of the troublesome courtship between the Thames and the Medway are only briefly hinted at and then are left behind; the major emphasis is given to the celebration of their union. The wedding feast is held, not surprisingly, in "*Proteus* house"; the story proceeds quite smoothly: it is all ceremony, a ritualized pageant, a procession in which

All which long sundred, doe at last accord
To joyne in one, ere to the sea they come,
So flowing all from one, all one at last become.
(xi.43.7–9)

The union is achieved naturally; original harmony is restored easily, gracefully, without effort or trouble.

The narrator's job, as he presents it, has been relatively easy as well. To relate the story properly and as completely as possible, he has requested the help of the "records of antiquitie," which "no wit of man may comen neare," appealing to the Muse to whom those records "appeare" (xi.10.4–5). The story of this natural union is grounded in sanctioned texts to which men ordinarily have no access; aided by the revelation of them that he is granted through the intercession of the Muse (though it does not resolve all his difficulties), the narrator, now self-proclaimed spectator and transcriber, recounts ("reherses") the procession and records the harmonious marriage.[36]

This incident of undisturbed union—or reunion—clearly stands in opposition to the more troublesome, as-yet-unresolved, and still potentially unresolvable story of Florimell and Marinell, for which there is, apparently, no recourse to these or similar sources and none of their sanction; it is to this more problematic episode that the narrator next forces himself to return. In the last stanza of Canto xi, the last stanza devoted to the rivers, we are told of a guest present at the banquet—Cymoent (or, as she is called here, Cymodoce), Marinell's mother. But the narrator, having taken a rather circuitous route to reach the end of the Florimell and Marinell episode, is weary, as is his Muse ("her selfe now tyred has" [xi.53.8]), and—to refresh himself after his previous endeavor and to prepare himself for the more difficult one that lies ahead—he pauses now to catch his breath, passing to the next canto to reenter fully the world of the characters who inhabit his Faeryland. There we discover that Marinell has accompanied his mother to the wedding, to "learne and see / The manner of the Gods when they at banquet be" (xii.3.8–9). This perhaps suggests that, although no "antique" account of his story exists, the Thames and Medway story of ideal union can serve as a model for his activities, but we soon learn that Marinell, being "halfe-mortall" (unlike his sea-nymph mother, he is not "devoyd of mortall slime" [III.iv.35.3]), cannot participate in the ceremony; he too is a spectator of this scene and ultimately does not have complete and free access to it: "He might not with immortall food be fed, / Ne with th'eternall Gods to bancket come" (IV.xii.4.3–4).[37] Marinell's situation, imposed upon him by his mortal nature—by his origins—now mirrors the narrator's in relation to Florimell; for when, as he wanders around "*Proteus* house," he hears Florimell's complaints and learns of her plight, he finds himself equally helpless, equally unable to do anything to aid her. His resistance to her vanishes, but he is completely impotent; he can only, like the narrator in Canto xi,

> . . . inly wish, that in his powre it weare
> Her to redress: but since he meanes found none
> He could no more but her great misery bemone.
> (xii.12.7–9)

And now he shares the poet's sense of responsibility, to which he responds with a similar style of self-castigation: after trying to devise plans to save Florimell, he realizes that they are all futile, and

> At last when as not meanes he could invent,
> Backe to him selfe he gan returne the blame,
> That was the author of her punishment.
> (xii.16.1–3)

Cymoent, returning to her son to discover him near death because of all this, is also unequal to the task; she even has difficulty discerning what the problem *is*. Apollo finally reveals the details to her; then, knowing that any appeal to Proteus (who is "the root and worker of her woe" [xii.29.2]) will be in vain, she appeals to "King *Neptune*," who, we learn, supersedes Proteus's authority—an authority that has, till now, been unquestioned, though vile. In fact, we learn that Proteus's power over Florimell is a temporary fiction; the control he has been claiming and executing is not really his. As Cymoent explains to Neptune, Proteus has acquired Florimell by an act of "unjust possession":

> For that a waift, the which by fortune came
> Upon your seas, he claym'd as propertie:
> And yet nor his, nor his in equitie,
> But yours the waift by high prerogative.
> (xii.31.3–6)

His control is therefore fragile, unstable, incomplete, unauthorized. To be sure, his rights of possession are now quickly dispelled, the thing unjustly possessed now quickly retrieved. For Neptune "streight his warrant made, / Under the Sea-gods seale autenticall, / Commaunding *Proteus* straight t'enlarge the mayd" (xii.32.1–3), and Proteus, though "grieved to restore the pledge he did possesse" (xii.32.9), knows that the game is over, and "durst he not the warrant to withstand" (xii.33.1). Florimell is freed and delivered to Marinell. But even this union is incomplete; the divine fiat of Neptune (the miracle hoped for by the narrator in the beginning of Canto xi) does not resolve all difficulties. There is no marriage, and no celebration; the narrator still cannot complete the story, which, he concludes, "to another place I leave to be perfected" (xii.35.9).

The relationship between Faeryland[38] and the natural world in Book IV is almost the exact opposite of the relationship between the Faery domain and the human world presented in the chronicles Guyon and Arthur read in

Book II. There, the Faery domain is for the most part peaceful and orderly, with its peace and order naturally obtained through lawful succession; the disruptive essence is only briefly suggested. Human history, on the other hand, is mostly disruptive and difficult. Here, Faeryland is full of usurpations, betrayals, and disorder, while natural history, despite a troublesome beginning (the courtship) that is only hinted at, proceeds peacefully and gracefully, without interruptions. This pattern reinforces the notion of the continuum that, I suggested, contains both the idealized vision created in Faeryland and its disturbing reality as enacted by man, for now Faeryland reveals, as its main attributes, all the negative features formerly found in human history, and the ideal is outside of its sphere. In fact, Faeryland and the human world have been conflated here: if Marinell cannot fully partake of the ideal world of the rivers because he is *half*-mortal, then full mortals would, one supposes, like the narrator, have similar, if not even more severe, restraints imposed upon them. The possibility that Faeryland—now practically identified with the world of mortals—can embody the idealized vision is subverted; like Proteus's power to acquire and retain his unlawful possession, Faeryland's potential—as exhibited in Book II—truly to possess its stolen gift of ideal life is revealed as an illusion and is overruled.[39] The consequences of that initial act of creation have been realized; the thing created, unauthorized, has been recalled. The ideal is now portrayed as existing outside the world of human endeavors and also—because they are now seen as identical—beyond the reach of the poet's fiction, beyond the scope of his created world. The narrator, it seems, is confronting his inability to capture that ideal state, which here stands in ironic contrast to—and as inaccessible to—man and, more particularly, to the poet and his fiction. The ideal cannot be legitimately and fully located in the fictional world he creates, which is now recognized as a world that resembles rather than reverses the actual world; it remains remote, in the world of nature, a final vision of peace and harmony ultimately denied to man and his fictions, to ordinary humans as well as to poets.

The process implied by the continuum has occurred; the ideal of Faeryland has receded. The poet has no authority to create in his imagined world the perfected vision; his text instead closely adheres to the actuality he so wishes to transcend. Like Marinell, who can watch the harmony of the gods but can never enact it or incorporate it into his own life, the poet can read about and record harmony but cannot enact it or model his Faeryland on it. The known world is his model; it structures his text. The golden world that can be evoked in Faeryland can be no more permanently inscribed than the ideal visions that any mortals (or, for that matter, faeries) who inhabit it can ever see or achieve: no more enduring than Red Cross Knight's vision on the Hill of Contemplation, no less fleeting than Arthur's vision of Gloriana in his dream.[40] In the Faeryland created by a poet who is tied to—and bound by the limitations of—"what is," ideals can be glimpsed but not finally embodied, momentarily grasped but never fully possessed or enacted.

Commentators have noted that in the last three books of *The Faerie Queene* there is a growing self-consciousness and disillusionment as the poet despairs about the efficacy of his poetry; the narrative begins to attend explicitly to its own composition and to the issue of writing itself. The problem, as it is presented by most critics, is that the poet becomes increasingly aware that the actual world is antagonistic to his Faeryland, and he comes to realize that his poetry cannot deal with this corrupt contemporary reality. Roger Sale's analysis is the most topical: he explains that in response to political developments in England and Ireland, "Faerie Land has collapsed."[41] Harry

Berger's explanation is even more representative: according to him, beginning with Book IV, "Faery comes into sharp conflict with the demands of the actual world"; there is a "growing loss of faith in history. . . . The poet cannot find the traditional and expected patterns of order in the chaotic world that surrounds him"; and "the poet is no longer certain about the ability of his imaginary forms to deal with the facts of social existence."[42]

The problem with these statements is their failure to locate correctly a major source of the poet's anxiety. To claim that either newly developed conditions in the actual world or the poet's newly developed awareness of the condition of that world causes the disintegration of his ideal Faeryland is to ignore the whole basis of allegorical writing. Inherent in allegory is the recognition that the world is corrupt and fallen; this concept forms the foundation of the mode. What the poet confronts directly by Book IV is not merely that the world is fallen and disparate from the ideal, but that Faeryland cannot be kept detached from that world, and is itself disparate from the ideal. Here he presents all the disturbing elements that were left in the background in Book II: his poetic world is not autonomous, and he, like Marinell, has no authority to fashion in it a vision of peace and harmony; like Proteus, he can make no final claims to the authority he formerly professed. The allegorical mode that could create the ideal Faeryland of Book II also insists that Faeryland ultimately is part of the actual world where those ideals cannot be permanently possessed, a world from which those ideals recede as soon as they draw near. The poet does not see conflict between Faeryland and actuality, he sees congruence; he does not so much acknowledge that "real life" contradicts his art as that it defines and structures it. He does not lose confidence in his ability to "deal with the facts of social existence"; he loses confidence in his ability to create a golden world independent of those facts—

indeed, here he illustrates his inability *not* to deal with those facts. The poem does not suddenly change because it must admit a disturbing reality; it demonstrates the implications of its own nature, whose reliance on and participation in that disturbing reality have already been adumbrated.[43] The successive books of *The Faerie Queene* do not "grow increasingly tenuous and inconclusive";[44] rather, the poet realizes and acknowledges that they have been, all along, tenuous and inconclusive. The poet comes to terms not simply with the state of the world—something he has always been aware of—but with his own mode of expression, which subscribes to the actual, betrays his lack of autonomy, and reveals his inability to fashion and authorize ideal resolutions that other sources can produce.

III

The *Mutability Cantos* of *The Faerie Queene*, like the "man of gret auctorite" who appears at the end of Chaucer's *House of Fame*, stand as the conclusion of a seemingly incomplete work, and they likewise exhibit an inconclusive, only temporarily final quality. As the opening allusions to the "records permanent" of Faeryland suggest, the issues raised in Book II and their consequences enacted in Book IV also emerge in the *Mutability Cantos*, though here they are presented in a more dramatized form.[45] The story is about an attempted usurpation, about the difficulty of negotiating the claims over what each antagonist views as the other's "unjust possession," and about the nature and accessibility of a region that purports to be stable, orderly, and harmonious. The overriding concern is with external models of authority that represent permanent, ideal visions; with the narrator's relation to his own poem, the actual world, and those preauthorized visions of harmony; and with the status of the speaker's own authorship. The story records the poet's doubts concerning the possibility of relying on sources of authority outside of himself and the efficacy of insisting upon his own authority.[46] The *Cantos* enact a search for a truly authoritative vision and voice that can formulate and sustain the ideal, but that search is informed by the speaker's ambivalence and uneasiness both about other sources and about his own form of expression.

This attempt to establish a reliable and accessible voice of authority occurs on several levels, and it is carried out by various characters as well as by the narrator in the *Mutability Cantos*. If Mutability represents the actuality of earthly existence, her forceful presence reflects the pressure that world exerts as she claims to be the controlling and defining feature of all other domains. The other main characters—Jove, Cynthia, and finally (though in a different category) Nature—are the supposed representatives of the ideal realms divorced from the corrupted and debased world. The narrator records the confrontation of these characters, relating the story of their struggle for power and position, but also in the process relating the story of his own struggle as poet throughout *The Faerie Queene* and the final dilemma with which he is faced. Those characters initially set up as authoritative in the *Mutability Cantos* become, to a certain extent, figures of the narrator, and they reveal what has happened to his sense of himself as creator, possessor, controller, and authorizer of his own ideal golden world.

In Canto vi, first Cynthia and then Jove seem at the outset to occupy authoritative status that is untouched by the "cruell sports" of the adversary Mutability, who, when introduced into the poem, has already conquered "the face of earthly things" (VII.vi.5.1). But gradually, it becomes difficult to differentiate between Mutability and these otherworldly figures.[47] Cynthia, whom Mutability first challenges, is introduced as the ultimate source of stability, inhabiting the place most distant from the already "perverted" worldly frame; she "raignes in everlasting glory," and Mutability must climb to "the highest stage" of the empire of the heavens to confront her (vi.8). Her description emphasizes her permanence and priority and implies an authority whose credentials are invulnerable to and detached from the threats of alteration posed by Mutability. Although it is a long climb, however, Mutability has no trouble reaching Cynthia's palace; she encounters no obstacles and seems to have easy access to this realm. And once she appears before Cynthia, face to face, the two become less easily distinguishable. The terms of their descriptions merge: Mutability "boldly" issues her challenge (a description that appears appropriately to describe the pose of the aggressive adversary), but Cynthia responds equally "boldly," ignoring the threat and commanding Mutability to withdraw immediately—a command that Mutability "boldly" ignores, continuing to make her claims, even increasing their forcefulness (vi.11.1, 12.7, 13.2). Along the same lines, Cynthia greets Mutability's opening attempt at usurpation with a "sterne countenaunce," which is countered, in the next stanza, with Mutability's own "sterne looke" (vi.12.5, 13.9). If Cynthia's appearance and temperament tend to resemble Mutability's here, by the end of the canto the two characters share—or have exchanged—even more crucial characteristics. Mutability, who had initially been held responsible for perverting and altering the "meet order" of earth's "faire frame" and making it "accurst" (vi.5), at one point expresses dismay about the fallen state of men—"whose fall she did bemone" (vi.11.5). And in the Faunus episode, Cynthia enters the sublunary world, depicted as Diana, whose "heavy haplesse curse" (vi.55.3) alters and defaces the pastoral, paradisiacal setting at Arlo Hill. The change in the landscape, we are here told, is produced not by Mutability but by a now fragmented Cynthia-Diana figure who participates in—and even causes—the debased "face of earthly things."

As Mutability and Cynthia meet and mingle, little seems to remain of the latter's distinct and authoritative position; in fact, she is rendered inert by her opponent. The "heavenly crew," unaware of the confrontation but disturbed by the "suddaine lack of light" (vi.14.3, 15.5), turn for help in interpreting and controlling this new development to a heretofore unacknowledged power higher than Cynthia—one even more distant from the earthly realm. Behaving like a humanized, rabble crowd, the heavenly petitioners "ran forth" to Jove "in haste," "beating at his gates full earnestly," clamoring for him to appear ("Gan call to him aloud with all their might"), and demanding a conclusive interpretation—'To know what meant that suddaine lack of light" (vi.15.3–5). Now Jove is presented as supreme (Cynthia's status as "highest" apparently no longer pertains as a new and more reliable authority is sought): he is the "king of Gods" (vi.14.9), and his palace is "fixt in heavens hight" (vi.15.2). But Jove presents a sorry picture of the Grand Interpreter. First of all, he has not been perceptive enough to notice anything unusual occurring, and he reacts to the situation only when he hears the questions and appeal of the gods. Even then, he can offer no solution to the problem, but merely reflects their anxiety and guesswork. The gods had proposed to themselves that "*Chaos* broken had his chaine" (vi.14.6), and Jove, "troubled much," expresses a similar uneasiness:

> Doubting least *Typhon* were againe uprear'd,
> Or other his old foes, that once him sorely fear'd.
> (vi.15.8–9)

All Jove manages to do finally is to delegate someone else (Mercury) to go and get the facts. Mutability's quick retort to Mercury, however, completely disregards and debunks Jove's and (once again) Cynthia's status as authorities: ". . . shee his *Jove* and him esteemed nought, / No more than *Cynthia's* selfe . . ." (vi.18.8–9). When Mercury returns with Mutability's reply, Jove learns that it is not Typhon but another of those old but still-unconquered foes who is challenging them again. His response is to review how the gods—like Calidore, who could temporarily capture but not eliminate the threat of the Blatant Beast—had failed to defeat completely "th'Earth's cursed seed" who had once before "Sought to assaile the heavens eternall towers":

> . . . we then defeated all their deed,
> Yee all doe knowe, and them destroied quite;
> Yet not so quite . . .
>
> (vi.20.5–7)

Even those farthest removed from the native ground of the representative of earthly disorder cannot quell or fully resist her power to infiltrate their domains.

This infiltration is again expressed through the description of the two opponents, who share terms and traits. When Jove hears of Mutability's challenge, his "count'-nance" remains "bold," and he then denounces "this bold woman" who "with bold presumption doth aspire" (vi.19.7, 21.1–2). He proceeds to turn the group of gods into a council, with himself in charge, outlining the details of Mutability's uprising and declaring that the issue must be resolved:

> Wherefore, it now behoves us to advise
> What way is best to drive her to retire.
>
> (vi.21.6–7)

At the same time, Mutability has been engaged in a similar process:

> Meane-while, th'Earth's daughter . . .
> . . . gan now advise,
> What course were best to take in this hot bold emprize.
> (vi.22.7–9)

Then again, at a crucial moment, expected characteristics are exchanged and reversed. Like the Goddess Nature in the *Parliament of Fowls,* who is unable to fulfill her authoritative position and transfers it onto the company of birds, Jove fails to perform his role as controlling leader and instead displaces the responsibility for articulating an authoritative judgment onto the "heavenly" but humanlike "crew": "Areed ye sonnes of God, as best ye can devise" (vi.21.9), and "So having said, he ceast" (vi.22.1). Contrarily, Mutability does not seem to suffer from the incapacities that plague Jove: once she has spelled out the problem, "Eftsoones she thus resolv'd" (vi.23.1). And when, immediately acting upon her forceful and determined decision, she arrives at Jove's palace, the gods are still "troubled, and amongst themselves at ods" (vi.23.3) and "ne wist what way to chose" (vi.24.5). The heavenly regions seem to embody all the disorder and inconclusiveness formerly assigned to earth, while authoritative control now seems to reside in earth's daughter, Mutability.

As they confront each other, Mutability and Jove become, once more, difficult to distinguish. To meet Mutability, Jove resurrects himself as the picture (if nothing else) of the ruling power. But even this serves to erase distinctions: when he "gan straight dispose / Himselfe more full of grace" (vi.24.7–8), Mutability, though "almost queld" by this performance, similarly pulls herself together, responding with a similar demeanor: all present "marked well her grace" (vi.28.2). Their modes of address and attack mirror each other as well: Jove attempts to deflate both the figure and her challenge ("Then ceasse thy idle claime thou foolish gerle" [vi.34.1]), while Mutability deploys similar tactics ("Ceasse *Saturnes* sonne, to seeke by proffers vaine / Of idle hopes t'allure mee to thy side" [vi.34.7–8]). As Jove diminishes Mutability's actions to the vain demands of a reckless child, she in turn lowers his to that same level—merely a son's foolish and ineffectual desires.

By the end of these two encounters, no hierarchy or criterion for differentiation remains. Whenever an attempt is made to raise a "higher" figure well above the earthly Mutability, or to diminish Mutability to a position far below it, they all become, instead, images of each other; no distance can be maintained. Although the characters and the narrator do not acknowledge it, there is a constant equalization occurring, which makes it difficult to distinguish the two realms, the opponents, and their claims. No matter what tactics are employed, Cynthia and Jove cannot remain detached from Mutability; however they act, they reveal their resemblance to rather than their separateness from her. It is, finally, impossible to assign any distinct characteristics to the gods that differentiate them from Mutability, and this suggests the futility of trying to uphold an authority that may be accessible to an earthly traveler but that, for that very reason, cannot be sustained.

Jove and Cynthia, in short, cannot maintain their authoritative positions as rulers of an ideal, stable region, nor can they deny association with the world of Mutability. In this way, they are like the narrator, who cannot authorize an ideal, harmonious Faeryland unconditioned by the fallen earthly world, for it inevitably reveals its resemblance to that world. The issues at stake for the characters in the *Mutability Cantos* mirror the issues at stake for the narrator throughout *The Faerie Queene*. The question is one of rightful possession: Mutability claims she has been unjustly deprived of a realm to which she feels entitled ("by the fathers . . . / I greater am in bloud (whereon I build) / Then all the Gods, though wrongfully from heaven exil'd" [vi.26.7–9]), and she contends that Jove is attempting to retain an "unjust possession" ("thou *Jove*, injuriously hast held / The Heavens rule from *Titans* sonnes by might" [vi.27.6–7]). As she later makes clear, both Cynthia and Jove, despite their status now, are no less tied to the earth than she—they are "mortall borne" (vii.50.5), "begotten were, / And borne here in this world" (vii.53.8–9). In response, Jove insists upon his present status and invokes the Promethean legend—now returned to its traditional context—against Mutability, to show her the illegitimacy of her own claims and the consequences of such presumptions: "I would have thought, that . . . // . . . great *Prometheus*, tasting of our ire, / Would have suffiz'd, the rest for to restraine" (vi.29.5–8). Mutability accuses Jove of desiring rule over a domain he has no real authority to possess or acquire; Mutability is presented as attempting to claim ownership over what she cannot justifiably call her own; and Prometheus now symbolizes not—as he did in Book II—the unauthorized act of creation that is able to produce a stable, ideal realm, but rather the inevitable demise of such futile, because earthbound, endeavors.

As it becomes increasingly difficult to make distinctions between, on the one hand, the region initially introduced as authoritative, stable, and orderly and, on the other hand, the perverted earthly landscape, the characters—none of whom can validate his or her rights to the heavenly realm—discover that they cannot negotiate their own claims. As when Cynthia's status was destroyed, now, after Jove's inadequacies are revealed, a new "highest" is uncovered who may be able to end the stalemate. In a statement that unequivocally equalizes Jove and herself, Mutability proposes that they bring their case to "the highest him, that is behight / Father of Gods and men by equall might; / To weet, the God of Nature" (vi.35.4–6)—and Jove grudgingly agrees, consenting to the notion of the appeal and implicitly conceding to the new hierarchy (which places him and Mutability on the same level, though still as adversaries) that the suggestion carries.

The disputing parties adjourn to Arlo Hill, where Nature holds court, and all, including the narrator, testify to her special position as "highest":

> Then forth issewed (great goddesse) great dame *Nature*,
> With goodly port and gracious Majesty;
> Being far greater and more tall of stature
> Then any of the gods or Powers on hie.
>
> (vii.5.1–4)

Nature exists on a level different from that of all others present at the council, and the narrator's attempt to describe her reveals the problematic nature of her authoritative position and its relation to her audience, of which he is a part.[48] The difficulty here is not (as it was with Jove and Cynthia) that she fails to act as an authoritative figure who embodies ideal forms, but rather that her status as such is inaccessible to and conflicts with the individual, more circumscribed perspective of her audience. Nature's primary characteristic is that she incorporates seeming opposites:

> This great Grandmother of all creatures bred
> Great *Nature*, ever young yet full of eld,
> Still mooving, yet unmoved from her sted;
> Unseene of any, yet of all beheld.
>
> (vii.13.1–4)

Processes and qualities ordinarily considered contrary are reconciled in—and by—Nature; the goddess harmonizes elements that from a different, more restricted viewpoint seem discordant, even mutually exclusive.[49] Because of this, however, Nature presents problems of interpretation for her audience, none of whom, for example, could "well descry" "Whether she man or woman inly were" (vii.5).

In other words, the narrator's description of the reaction to Nature's appearance demonstrates that those positioned below Nature regard her through an either/or dichotomy and are unable to comprehend her capacity to embody opposites. Furthermore, rather than provoking an harmonious reaction, she produces alternative and discordant judgments.

These problems of comprehension are coupled with problems of expression. Nature's ability to harmonize contraries runs up against linguistic barriers, for as the narrator's description of her indicates, this harmony can be portrayed only through terms that necessarily fragment it once again into opposites. Even the goddess's mode of self-presentation is an occasion for concern, not celebration. In Macrobian fashion, she is covered with a veil, which, like the veil of allegory, is at first described as a device that hides and conceals the figure from view:

> . . . with a veile that wimpled every where,
> Her head and face was hid, that mote to none appeare.
> (vii.5.8–9)

But two stanzas later, the narrator has decided that the purpose of the veil is not really to conceal; he subscribes to the interpretation that "others tell":

> . . . [Nature's face] so beautious was,
> And round about such beames of splendor threw,
> That it the Sunne a thousand times did pass,
> Ne could be seene, but like an image in a glass.
> (vii.6.6–9)

In other words, Nature is veiled not to prevent others from seeing her but to enable them to do so. The more lowly creatures are not equipped to view Nature's image directly, and the only way she can reveal herself to them is through the deflecting, covering robe.

But although this description of Nature's veil suggests

its similarity to the veils used by allegorical poets, Nature's is different in one crucial respect—it does not distort the figure to accommodate the perception of her viewers. The garment may mute the force of the goddess's unique qualities so they do not blind her audience, but it does not change their essential form; it remains faithful to the character of Nature instead of adjusting itself to that of her audience. Thus the veil has all the properties of the figure it covers, and it re-creates, rather than dismisses, the same problems for the writer:

> Her garment was so bright and wondrous sheene,
> That my fraile wit cannot devise to what
> It to compare . . .
>
> (vii.7.3–5)

Nature's face, which throws off such beams of splendor that she cannot be fully gazed upon, is covered by a veil that, shining with similarly stunning rays, cannot be fully described; as the poet admits, no analogies in his language are adequate to represent it.

But while human art is unable to respond correctly to Nature, the natural world itself has little difficulty:

> In a fayre Plaine upon an equall Hill,
> She placed was in a pavilion;
> Not such as Craftes-men by their idle skill
> Are wont for Princes states to fashion:
> But th'earth her self of her owne motion,
> Out of her fruitful bosome made to growe
> Most dainty trees; that, shooting up anon
> Did seem to bow their bloosming heads full lowe,
> For homage unto her, and like a throne did shew.
>
> (vii.8)

Human art falls short of nature's power to convey and accommodate the figure of Nature, and comparisons always reflect badly on the human artificer: even the flowers underneath Nature's feet

> . . . richer seem'd then any tapestry,
> That Princes bowres adorne with painted imagery.
> (vii.10.8–9)

The natural world seems to have an easy relationship with the goddess that is not shared by men or artists.

These shortcomings of human response and expression provoke the narrator to reflect on his own status as storyteller and artist. Conceding his inability to describe fully what Nature both presents and represents, he traces the history of this problem—and displaces it by referring to Chaucer and Alain de Lille:

> So hard it is for any living wight,
> All her array and vestiments to tell,
> That old *Dan Geffrey* (in whose gentle spright
> The pure well head of Poesie did dwell)
> In his *Foules parley* durst not with it mel,
> But it transferd to *Alane,* who he thought
> Had in his *Plaint of kindes* describ'd it well:
> Which who will read set forth so as it ought,
> Go seek he out that *Alane* where he may be sought.
> (vii.9)

Citing Chaucer's problems, the narrator also cites Chaucer's solution: locate a copy of Alain de Lille's *De Planctu Naturae*. But Chaucer never resolved the issue of Nature's problematic status as an authority, and even Alain realized the limits of communication between the goddess and man: the human audience in his poem can neither interpret nor even recognize Nature, whose veil has been torn and rent by the disruptive activities of men who pervert her standards.[50] The entire issue of how best to respond to and describe Nature is, in the *Cantos,* quite muddled; though she stands as the final arbiter and authority, she cannot be rendered in terms that her audience can easily assimilate. The narrator devotes nine stanzas to Nature only to point once again to the problems involved in trying to understand and represent her.

These characteristics of Nature and her relation to her audience explain both the content and the effect of her final verdict after she hears the appeals of Jove and Mutability. Nature first acknowledges the mutable, unstable quality of the world:

> I well consider all that ye have sayd,
> And find that all things stedfastnes doe hate
> And changed be . . .
>
> (vii.58.1–3)

But she also explains that the mutable and the unchangeable cannot be separated, for this admittedly unstable world is the medium through which stability and perfection are formed:[51]

> . . .yet being rightly wayd
> They are not changed from their first estate;
> But by their change their being doe dilate:
> And turning to themselves at length againe,
> Doe worke their owne perfection so by fate:
> Then over them Change doth not rule and raigne;
> But they raigne over change, and doe their states maintaine.
>
> (vii.58.3–9)

Nature here consolidates and articulates what was implicit but not fully acknowledged earlier in the *Cantos*—that clear-cut distinctions between Jove's realm and Mutability's cannot be made, since stability must manifest itself and work itself out through the mutable and earthly.[52] She combines and reconciles into a single vision what the limited, either/or perspective of Jove, all the gods, and even the narrator persisted in viewing as contrasted: although previously the adversaries and what they represented were constantly becoming indistinguishable, they were still considered—and considered themselves—as warring factions. Nature's all-inclusive vision dismisses the strict polarities and creates an ideal harmony between them. Moreover, in the process of defining the relationship that pertains between Jove and Mutability, Nature seems to provide a capsule description—from a positive perspective—of what the narrator has discovered and found disturbing about his own poetry: just as Jove's domain of stable order must take shape through Mutability's domain of change and alteration, so the poet's ideal visions cannot subsist autonomously and must be expressed through the form of the actual.

There is, however, a crucial difference that explains Nature's favorable attitude toward and understanding of this interdependent process, as well as the narrator's failure to share it. Like her own mode of presentation, which uses a veil that does not compromise her essence, perfection, according to the goddess, is always the presiding force, and is not compromised by the form it takes on earth. The ideal state may need to work itself out through the unstable and fickle earthly world, but for Nature this requirement does not, in the end, distort the essence of that perfection, because the end *is* that perfection:

> But time shall come that all shall changed bee,
> And from thenceforth, none no more change shall see.
> (vii.59.4–5)

Nature's speech concludes with this image of a final, permanent ideal state.

But man's understanding and ability do not mirror Nature's, and it is not Nature's judgment that concludes the *Cantos*. The earlier stanzas that introduced Nature indicated the problems that Nature's vision, perspective, and function created for her audience: her availability to human comprehension and expression was questioned and found incomplete. Here, too, the force of Nature's decree is qualified; despite all her status as an authority, she does not have the final say for her audience. Immediately after she speaks, gaps of authority and interpretation are created. The figure whose judgment has been posited as truly authoritative "did vanish," and the phrase that follows her disappearance—"whither no man wist" (vii.59.9)—highlights the shortcomings of her audience: certain kinds of knowledge and forms of expression remain beyond man's grasp.

The narrator rushes to fill these gaps—and Nature's role—in the next (and last) canto, by reviewing and deliberating the debate and Nature's verdict himself. However, although when Nature vanishes the narrator assumes her position as final arbiter and interpreter, he does not deploy her interpretive principles: he does not articulate her harmonious unification, nor can he attain and sustain her final ideal of perfection.[53] Rather than reasserting resolution, the poet re-creates discord, which is even imaged in the structure of his two-stanza conclusion. Contrary to Nature's argument that although "all things" change, they are not ruled by Change but rather "raigne over change," the narrator's perspective in the first stanza echoes Mutability's and his own as expressed earlier in the *Cantos*. Revealing his dependence on and ties to the evidence of visible, external, actual "things" that are "so fading and so fickle," he concludes that in the present, on earth, Mutability "beares the greatest sway" (viii.1); everything is unstable, everything decays. Then, turning in the next stanza to reconsider what Nature has said, the narrator ignores the first part of the goddess's response—where she reconciles the orders of the mutable and the unchangeable—and he speaks only of her final words, her prefiguration of perfection:

> Then gin I thinke on that which Nature sayd,
> Of that same time when no more *Change* shall be,
> But stedfast rest of all things firmely stayd

> Upon the pillours of Eternity,
> That is contrayr to *Mutabilitie.*
>
> (viii.2.1–5)

But the narrator has merely reproduced Mutability's reading of change in the present and set it against a misreading of Nature's explanation of her final vision: whereas the goddess expressed, to borrow a term from the Garden of Adonis, "eterne in mutabilitie" (III.vi.47.5), the narrator interprets permanence and perfection as opposed to change and the earthly—he sees only an "Eternity, / That is contrayr to *Mutabilitie.*" The narrator cannot affirm the harmonizing that, from the perspective of Nature, is viewed positively as victory; he defines the ideal as antithetical to and detached from the earthly (since for him, to reconcile them is to lose the former), and in so doing he reveals that the ideal is beyond the grasp of his own poetic narrative. As the narrator speaks, in other words, and as his voice replaces Nature's, we witness, once again, the limits of his mode of expression: the first stanza betrays a voice that is formed and informed by an earthbound perspective, a voice that therefore (as the second stanza confirms) cannot articulate the ideal, which, consequently, must be viewed as existing outside of its domain.

The narrator thus concludes with an expression of his desire for—but not his attainment of—the goddess's vision:[54]

> But thence-forth all shall rest eternally
> With Him that is the God of Sabbaoth hight:
> O! that great Sabbaoth God, grant me that Sabaoths sight.
>
> (VII.viii.2.7–9)

The narrator appeals for an all-inclusive, truly authoritative picture that will release him from the discordant picture he himself presents. While Nature sees perfection working through but never threatened by a fallen world over which it always has priority, the narrator sees the world of Mutability as the ruling force during that period of interpenetration. And while Nature presents the ultimate perfection as a domain that is autonomous because it completely subsumes the influence of the mutable world, the narrator cannot sustain that picture; it disintegrates even as he speaks, for he represents the perfect state, even when it is achieved, as forever in conflict with and ultimately exclusive of an antagonistic mutable world—the world that shapes and defines his own pronouncements. Nature's ideal has receded along with the figure who voiced it, and when the narrator steps in to recapture it, he invokes all the problems she solved and can only pray for the ability to reinscribe the vision that has been and still is inaccessible to him. The poet cannot provide visions of those autonomous states of perfection himself; he simply asks for those moments when he is granted sight of them.

In these last lines, the narrator expresses his desire to "rest eternally." But rest—a word used twice in the last stanza and implied in one meaning of *Sabbaoth*—is a condition that has no real existence or status in Faeryland; it is a state usually coupled in *The Faerie Queene* with the recognition that it is impossible to achieve. From Book I through Book VI, no resolution or conclusion is final or complete—climaxes always recede and simply release new attempts to achieve them.[55] Red Cross receives his vision from the Hill of Contemplation but then must leave it to return to his quest; later, harmony is restored at the court of Una's parents only to be disrupted by the reentry of Archimago and Duessa; and after Red Cross is reunited with Una, he must return to the Faerie Queene's court. The Bower of Bliss is destroyed, but Grill remains Grill; Britomart's climactic rescue of Amoret is in part defeated because Scudamour's absence prevents the final reunion; Artegall is united with Britomart but then must depart to finish his quest, which remains incomplete; Calidore sees

Colin Clout's vision of the graces, but it vanishes as he watches, and he finally captures the Blatant Beast only to have it escape once again in the final stanzas of Book VI.

Several critics have noted that the most attractive—but also the most dangerous—temptation in *The Faerie Queene* is to rest: to cease the necessary activity of constantly attempting (in the face of only temporary achievements) to attain final resolutions.[56] The exemplary Arthur rejects the notion of rest: after he is granted a vision of Gloriana in a dream, he awakens determined "From that day forth . . . / To seeke her out with labour, and long tyne, / And never vow to rest, till her I find" (I.ix.15.6–8). Rest is so dangerous precisely because it is so illusory: it implies a sense that one has fully achieved those conclusive states of perfection. "Eternall rest" is ostensibly the temptation that Despair offers to Red Cross, but what finally evolves in this scene is the delusive nature and the ultimate unavailability of the temper's offer—for Despair himself is unable to achieve that state of final repose:

> He chose an halter from among the rest,
> And with it hung himselfe, unbid, unblest.
> But death he could not worke himselfe thereby;
> For thousand times he so himselfe had drest,
> Yet nathelesse it could not doe him die,
> Till he should die his last, that is externally.
> (I.ix.54.4–9)

The fulfillment of that sustained final rest can be foreshadowed but never embodied in the world of change and mutability—the world in which the characters in *The Faerie Queene*, as well as the narrator himself, function and act.

In this context, the narrator's appeal for rest at the end of the *Mutability Cantos* expresses his disillusionment with, and his impulse to escape from, this process of continual striving and inevitable failure that defines his allegory and his Faeryland—the process that the poet's own voice creates when he reviews and rewrites Nature's judgment. This voice is not autonomous: it is bound to the concrete, external evidence of "things" that are "so fading and so fickle," and consequently it cannot authorize or articulate ideals that are unconditioned and uncompromised by the actual, visible, "knowable" state of things. Yet even as the narrator prays for a release from the conditions of his own poetry, he also regenerates them.[57] When the narrative voice enters and disturbs Nature's vision of final coherence, it reestablishes the pattern of the entire *Faerie Queene* and resurrects the continuum that structures his allegory: tied to the vicissitudes of "what is," the poet's Faeryland does not exist as an independent ideal realm; states of perfection are usurped possessions that the poet has no authority to maintain permanently. His mode of expression only permits movement on that continuum: movement toward ideals that soon become remote and that must be (unsuccessfully) striven for once more. The narrator's response to Nature thus reflects the level of action portrayed throughout *The Faerie Queene*: conclusions must be constantly worked toward, glimpsed, lost, and then worked toward again. Consequently, had Nature's verdict been allowed to stand as the final word, the poetic voice would have been surrendered; had her verdict not disappeared, Faeryland would have. By rewriting in his own voice Nature's final decree, the poet, though forsaking the ideal, ensures the possibility that his Faeryland can continue to exist, and he creates, once more, in his own terms—the only ones available to him—the possibility of momentarily viewing the ideal again. In short, he shatters (and loses) Nature's authoritative vision, but by doing so, he makes room for himself and his poem.

At the end of the *Cantos,* as the authoritative and autonomous Nature vanishes, a blank, quiet space is created. After the short silence between cantos, the narrator comes forth to fill that space. The moment he speaks, he reasserts his authorial status as final arbiter and interpreter, at the same time betraying his inability to fulfill the authoritative position left open by Nature's disappearance. He cannot endorse or faithfully represent Nature's judgment, but neither can he endorse or fully accept his own authorial voice, which disrupts the ideal. If the *House of Fame* concludes by breaking off the search for authority that has structured it, the *Mutability Cantos* conclude by renewing that search and that structure. If the *House of Fame* ends with the silencing of the authorial voice in order to sustain the image of an authoritative figure, *The Faerie Queene* ends with the resurrection of an authorial voice that simultaneously destroys and replaces the authoritative image and resumes the attempt to capture it. In his final lines—his plea for the vision that Nature so easily articulated and that he so immediately dispels—the narrator of *The Faerie Queene* indicates his uneasiness with, and his desire to be released from, but his refusal to relinquish, his own authorial voice.

CHAPTER IV

Creative Imitation in the Sonnets

I

In his letter to Boccaccio concerning the principles of literary imitation, Petrarch details the proper relationship between a model and its descendants. "A proper imitator," he advises,

> should take care that what he writes resembles the original without reproducing it. The resemblance should not be that of a portrait to the sitter—in that case the closer the likeness is the better—but it should be the resemblance of a son to his father. . . . As soon as we see the son, he recalls the father to us, although if we should measure every feature we should find them all different. But there is a mysterious something there that has this power.[1]

For Petrarch, then, imitation, while based on resemblance, is characterized by difference; or, as he writes, "with a basis of similarity there should be many dissimilarities." But the "mysterious something" that defines and identifies imitation seems forever elusive. "The quality is to be felt rather than defined," he adds.[2] As Thomas Greene notes in his commentary on this passage, "Petrarch is describing an object of knowledge that . . . cannot by definition be fully and succinctly delimited. The resemblance of a poem to its model or series of models will never be fully articulated, even supposing that it will be fully grasped."[3] As crucial as the question of whether this concept can be "articulated" or "grasped" is the question of whether it can be implemented. The anecdote that follows this discussion of imitation in Petrarch's letter radically qualifies the program for writing that he, however evasively, proposes. Although he had himself been "extremely punctilious, in spite of the difficulty, to eliminate all repetitions of my own words, and still more all echoes of previous writers," Petrarch relates how he was "struck dumb" to learn from his young disciple that he had, in his *Bucolicum Carmen,* incorporated verbatim a phrase from Virgil. The lesson to be learned, he concludes, is that "such things may happen not only to me, a conscientious scholar but unlearned and barren of wit, but to any man, however erudite, however abreast of our literature." If this comment suggests the extreme difficulty of ever fully following the prescribed method for imitation that makes "poets" rather than "apes," he goes on to suggest its impossibility: "For perfection is the property of him alone from whom we derive all our knowledge, all our possibilities."[4] The implication is severe. The ideal of true imitation can be neither fully articulated nor fully achieved. The true "poet," despite good intentions, gives way to the "ape" of reproductive techniques, the counterpart of which is the stupefied silence that accompanies the recognition that the words once regarded as one's own are, in fact, adopted, bearing the signature of another author.

The possibility of distinguishing between what one may call one's own (that of which one may legitimately be called author) and what retains its ties to a source outside of the author was a paramount concern for Petrarch. Despite his famous doctrine (which he duly ascribes to Seneca and Horace) of creative imitation—"we should write as the bees make sweetness, not storing up the flowers but turning them into honey, thus making one thing of many various ones, but different and better"[5]—Petrarch takes care to admonish writers not to lose sight of the early stages of this process. Although Greene is certainly right to emphasize the importance of Petrarch's idea of the "capacity for absorption and assimilation on the part of the poet, a capacity for making one's own the external text in all its otherness," Petrarch's own comments on this process do not as unambiguously suggest its virtues, and he is more tentative not only about the possibility of "making one's own the external text" but also about the rights of possession and autonomy implied in that notion.[6] In an earlier letter to Boccaccio, Petrarch also refers, with sanction, to the (vaguely defined) style that "transform[s] them [other authors] honorably, as bees imitate by making a single honey from many various nectars,"[7] but another narration of personal experience exposes the limitations and dangers both of engaging in this activity and of too readily assuming the possibility of performing it. Petrarch recounts how he has read the masters, "Virgil, Horace, Livy, Cicero, not once but a thousand times. . . . I ate in the morning what I would digest in the evening; I swallowed as a boy what I would ruminate upon as a man." The stages that Greene calls "absorption and assimilation" follow: "These writings I have so thoroughly absorbed and fixed, not only in my memory but in my very marrow, these have become so much a part of myself, that even though I should never read them again they would cling to my spirit, deep-rooted in its inmost recesses." But this process may have another stage, one that is not condoned:

> But meanwhile I may well forget the author, since by long usage and possession I may adopt them and regard them

> as my own, and, bewildered by their mass, I may forget whose they are and even that they are others' work. This is what I was saying, that sometimes the most familiar things deceive us most. They recur perhaps to memory . . . and they seem to be not merely one's own thoughts but, remarkably indeed, actually new and original.[8]

A thin and tenuous line, if any, separates "honorable" absorption and assimilation from wholesale adoption and usurpation. It is easy, too easy, Petrarch implies, to be deceived into thinking that borrowed words are one's original creation. He takes great pains to warn aspiring writers (as well as himself) that those "most familiar things" that seem to be the original property of the writer are most likely to be confused with what the writer has read many times in other books. For Petrarch, the concept of "making one's own the external text in all its otherness" would have been, it seems, something to be feared and avoided as well as praised and pursued; it is both a sign of achievement and an indication that some error has created the *illusion* of the accomplishment.

The digestive imagery Petrarch uses to describe a writer's method of engaging the work of his predecessors—imagery that can be traced back to Seneca—became a familiar feature of sixteenth- and seventeenth-century discussions of imitation.[9] In fact, the ambivalence about the relationship between a writer and those who exert influence upon him, about the status of a text that (presumably) earns its independence by incorporating other texts, is evident in the variety of interpretations given to the digestive metaphor. For Petrarch, the devouring of other authors was the beginning of a process in which consumption could become too complete; it made the words of others seem too inseparably a part of the reader's "marrow" and "spirit." In his *Ciceronianus* (1528), Erasmus has Bulephorus use a similar figure to oppose the servile Ciceronian Nosoponus with a definition of proper—creative—imitation:

> . . . that which culls from all authors, and especially the most famous, what in each excels and accords with your own genius,—not just adding to your speech all the beautiful things that you find, but digesting them and making them your own, so that they may seem to have been born from your mind and not borrowed from others, and may breathe forth the vigor and strength of your nature, . . . so that your speech may not seem a patchwork, but a river flowing forth from the fount of your heart.[10]

Although arguing, as Petrarch does, that material should be culled from many texts (rather than a single text) and accomodated to the individual character and style of the writer, Erasmus unequivocally endows the digestive process with positive qualities. Whereas Petrarch argued against the tendency to so completely (and almost unconsciously) digest other writings and "regard them as [one's] own" that "they seem to be not merely one's own thoughts but . . . new and original," Erasmus encourages this very activity of "digesting them and making them your own, so that they may seem to have been born from your mind and not borrowed from others." Erasmus is, of course, advocating Petrarch's concept of transforming other writings "honorably," not outright plagiarism, but in doing so he uses the same language that Petrarch employed for the opposite purpose. The process by which any text becomes so independent of its models that it can be called "new and original," by which "borrowed" words become so assimilated that they can no longer be said to belong in their original context, is barely differentiated from the process that deceives a writer into calling his own the material that is not assimilated, the words that have retained their separate identity. What actually constitutes an honorable transformation and the ways in which it can be distinguished from servile imitation are so unclear that both operations can be described almost identically.

Almost a century later, the same imagery is being employed, but in a more complex form that seems to acknowledge a need to distinguish between the different kinds of activities. Bacon's advice, that "Some books are to be tasted, others to be swallowed, and some few to be chewed and digested," clarifies, but not in detail.[11] A more explicit distinction is provided by Ben Jonson, who suggests that there are two separate methods of consuming a text—one to be avoided, one to be pursued:

> The third requisite in our *Poet*, or Maker, is *Imitation*, to bee able to convert the substance, or Riches of an other *Poet*, to his owne use. To make choise of one excellent man above the rest, and so to follow him, till he grow very *Hee*: or, so like him, as the Copie may be mistaken for the Principall. Not, as a Creature, that swallowes, what it takes in, crude, raw, or indigested; but, that feedes with an Appetite, and hath a Stomacke to concoct, divide, and turne all into nourishment. Not, to imitate servilely, as *Horace* saith, and catch at vices, for vertue: but to draw forth out of the best, and choisest flowers, with the Bee, and turne all into Honey, worke it into one relish, and savour: make our *Imitation* sweet: observe, how the best writers have imitated, and follow them.[12]

Jonson differentiates swallowing whole—simply incorporating, without performing any activity upon the ingested material—from the actual process of digestion that transforms the "raw" material into "nourishment" for the new body. Analyzed in this manner, the imagery of eating can distinguish between apish imitation, which simply copies a previous work and therefore never achieves autonomy, and creative imitation, which adapts the old material and therefore establishes a new and separate identity.

But another complicating feature is introduced by the imagery of Jonson's opening remarks.[13] When Petrarch insists that there should always be a recognizable difference between any single work and the model it draws upon, he (again following Seneca) argues against the analogy of a portrait, in which case "the closer the likeness is the better." Jonson's language, however, reverses the emphasis: he stresses resemblance, advocating that the poet become the image of his predecessor, that the copy become the image of the original ("so to follow him, till he grow very *Hee*: or so like him, as the Copie may be mistaken for the Principall"). Not only may descriptions of conflicting processes share the same images and terms, but, as indicated here, even when the proposed concept of imitation does not change (cf. the bee-and-honey analogy), the various descriptions of it can be totally contradictory. Even Jonson's final counsel, to "observe, how the best writers have imitated, and follow them," offers little concrete advice, for the notion that imitation itself must be imitated presupposes a clear and accurate guide among the many models. The earlier writers had already seen the problem in this assumption, and Pico's opening words in his letter to Bembo acknowledge the dilemma:

> I was in doubt, Bembo, whether I ought to agree or disagree with you not only in your imitation of ancient writers but also in your opinions on imitation, for I find that the ancients themselves who are proposed as worthy of imitation not only differ from one another in regard to this but also have changed their minds from time to time, and the arguments were so nearly equal on both sides that it was difficult to decide which had the advantage.[14]

And Castiglione describes Ludovico's echoing similar sentiments—expressing his confusion concerning those who discuss imitation—when he complains, at the court of Urbino, that:

> There are many who want to judge of styles and who talk about harmonies and imitation, but they are quite unable to explain to me what style and harmonies are, or in what imitation consists, or why things which are taken from Homer or from someone else are so proper in Virgil as to seem enhanced rather than imitated.[15]

As Pico and Castiglione suggest, the ambiguities are not only intertextual; they are also intratextual. Like Petrarch, who complicates his own program of creative imitation by declaring the inescapable potential for departing from it, Jonson seems to subvert his own stand on the relationship between model and copy. Although he does make the important distinction between kinds of consumption, advocating one that permits the imitator's own style and talents to mingle with and transform the ingested material, he also proposes that—at least for the beginner—slavish imitation (which, he implies, as Petrarch did, may be unavoidable for serious readers) can be a virtue, providing the writer with an authority he could not achieve on his own:

> . . . such as accustome themselves, and are familiar with the best Authors, shall ever and anon find somewhat of them in themselves, and in the expression of their minds, even when they feele it not, be able to utter something like theirs, which hath an Authority above their owne. Nay, sometimes it is the reward of a man's study, the praise of quoting another man fitly.[16]

Yet he can just as readily ridicule essayists like Montaigne, accusing them of indiscriminate and careless reading and borrowing because they

> in all they write, confesse stille what bookes they have read last; and therein their owne folly, so much, that they bring it to the *Stake* raw, and undigested.[17]

We can see, then, that Renaissance formulations of the theory of imitation—as a form of copying that transforms the model text (or texts) into one with a new and separate identity—characteristically manifest more of what Margaret Ferguson has called (referring to Du Bellay's *Deffence*) "a significant pattern of contradictions" than what William Kerrigan, for one, has described as a "happy poise" imbued with a sense of compatibility rather than strain.[18] Yet although modern criticism generally uses the phrase *creative imitation* to refer to this theory, the tendency has been to dismiss the ambiguities posed (presumably) by our modern definitions of the terms *creation* and *imitation* and to present a synthesized account of it—one that obscures the difficulties presented by the Renaissance writers themselves. And although we use—now as a commonplace—the term *paradoxical* to characterize the Renaissance view of things, it has become customary to employ even this notion to smooth out difficulties. Dorothy Connell, for example, describes the "characteristic of Renaissance thought" as "an ability to encompass and balance contradictions"; more specifically, Nancy Struever speaks of the "paradoxical" Humanist "insistence on reciprocity—of liberty as existent only within limitation, . . . tradition as the ground of innovation."[19] We can frequently observe in Renaissance texts, however, the inability to rest easy with and totally embrace these "paradoxes," or even consistently to formulate them so that they read as comfortable reciprocities.[20] In many cases, discomfort with these "paradoxical" conjunctions is revealed in qualifying remarks that accompany their proposal, in overt attempts to explain away contradictions that are clearly felt, or in the presentation of the two halves of the paradox as disjoint. Furthermore, we often witness the breakdown of those reciprocities in Renaissance texts: the writer's troubled awareness that during the process of imitation, the limitations may overwhelm the liberty, the traditions may take the place of (rather than promote) innovation; or, conversely, that the liberty may break the necessary bounds of the limitation, that innovation may ignore or abuse the traditions. Renaissance writers themselves seem unable to provide a coherent program for this elusive system of writing, and the theories proposed by individual writers not only jar with those proposed by others but are internally inconsistent as well.

Puttenham's pronouncement on this matter succinctly exposes the problems and contradictions facing the sixteenth-century poet:

> A Poet is as much to say as a maker. . . . the very Poet makes and contrives out of his owne braine, both the verse and matter of his poeme, and not by any foreine copie or example, as doth the translator, who therefore may well be sayd a versifier, but not a Poet. . . . And neverthelesse without any repugnancie at all a Poet may in some sort be said a follower or imitator, because he can expresse the true and lively of every thing is set before him, and which he taketh in hand to describe: and so in that respect is both a maker and a counterfaitor: and Poesie an art not only of making, but also of imitation.[21]

Isabel MacCaffrey calls this "the cheerful compromise of Puttenham," and Rosalie Colie comments that Puttenham "manages to accommodate the ambiguities of maker and imitator."[22] But the two concepts are brought together by juxtaposition, not by integration. On the one hand, the poetic material originates in the writer's "owne braine"; on the other, it is derived from that which "is set before him."[23] Puttenham's transitional phrase, "without any repugnancie at all," attempts to explain away the conflict simply by stating that it does not exist, but the comment clearly indicates the tension created by this juncture, and it draws attention to rather than dismisses the discrepancy between the two ideas. There is no "cheerful compromise": the accommodation is made at the expense of congruity.

Although Sidney is often presented as the proponent of the poet as maker, his *Apology* places him among those who would consider Puttenham's analysis not as a cheerful compromise but as merely a temporary coalition of ultimately unreconcilable ideas. Both the concepts expressed by Puttenham are put forward in (and were probably influenced by) the *Apology*, but not as congruent ones. In fact, there is a decisive split between the two notions in Sidney's work, unredeemed by any attempt either to account for their disparity or to reconcile their contradictions. O. B. Hardison's suggestion that "Close reading of Sidney's *Apology* leads inevitably . . . to the conclusion that it speaks in two distinct and discordant voices" is, I think, quite apt.[24] Sidney's statement that poets "doo meerely make to imitate" (p. 159) seems easily to pair the two ideas, unencumbered by any conflict, but his more detailed analyses of these concepts render this simple statement at best ambiguous. When Sidney attempts, near the beginning of the *Apology*, to isolate poetry from other arts, he differentiates the poet from those who "followe Nature" as well as from those (e.g., grammarians, rhetoricians, and logicians) who both provide and follow "artificial rules" (pp. 155–56). Counter to these other arts, which are bound by the restrictions imposed by their subjects, Sidney places poetry, for

> Onely the Poet, disdayning to be tied to any such subjection, lifted up with the vigor of his owne invention, dooth growe in effect another nature . . . not inclosed within the narrow warrant of her [Nature's] guifts, but freely ranging onely within the Zodiack of his owne wit.
>
> (p. 156)

The poet's only restrictions are those placed by himself; his limits are determined by his own capabilities, not by any external models. Indeed, "where as other Arts retaine themselves within their subject, and receive, as it were, their beeing from it, the Poet onely bringeth his owne stuffe" (p. 180). The historian, to whom Sidney often compares the poet, is "loden with old Mouse-eaten records, authorising himselfe (for the most part) upon other histories" (p. 162), but all the material that the historian is "bound to recite" the poet may "(if he list) with his imitation make his own"; he "citeth not authorities of other Histories," since he has "all, from *Dante* his heaven to hys hell, under the authoritie of his penne" (p. 169). The poet, then, eschews "artificial rules" and is not bound by Nature or other authors, nor does he acquire his authority from submission to anything external; he "bringeth his owne stuffe," and when he "make[s] to imitate" he "make[s] his own," all under the "authoritie of his penne."

All this is fine, a seemingly trouble-free account of creative imitation that grants the poet a domain he can identify as his own. But in the latter portion of the *Apology*, Sidney introduces a different notion of poetry and of imitation.[25] Here he "confesses," with the same tone of concession that was evident in Petrarch, that even "the highest flying wit" must "have a *Dedalus* to guide him." This Dedalus has "three wings": "that is, Arte, Imitation, and Exercise" (p. 195). Sidney's new definition of imitation is revealed in his complaint that writers ignore "artificiall rules" and "imitative patternes,"[26] particularly when we set this statement next to his earlier discussion of the poet's reliance on "his owne invention" rather than on the "artificial rules" that control the other arts. After introducing the Dedalean concept, Sidney proceeds to evaluate the works of other writers, bemoaning, at one point, the unfortunate defects "in the circumstances" of *Gorboduc* that make it unfit to "remaine as an exact model" (pp.

196–97). Imitation here, we can assume, refers now to careful imitation of other authors and to an ethic of verisimilitude, whereas earlier the emphasis was on the poet's "freely ranging onely within the Zodiack of his owne wit." As Hardison writes, "Instead of images of freedom and flight, emphasis is on control and guidance . . . By the same token, when the term imitation has been used, it has meant creation . . . Now, however, it is used in the rhetorical sense of 'copying the masterpieces.' "[27] However, while I would agree with Hardison's analysis, I cannot agree with his conclusion that of the two discordant theories presented in the *Apology*, the first "is the one that speaks the more effectively for the poetry of the Elizabethan period."[28] It is precisely the disparity between views, the unresolved tension between discordant ideas, that seems to be most representative of the period. Petrarch includes his own ambivalence as part of—and as qualification of—his discussion of creative imitation; although Sidney, like Jonson, simply allows the contradictions to infiltrate his essay without acknowleding them, Puttenham's need to assert that no conflicts exist in his treatise, which reflects the influence of the *Apology*, clearly indicates the uneasiness that such a coalition of ideas must have raised. The Renaissance writers themselves neither reveal a coherent system that synthesized the two concepts nor suggest that they would ally themselves with just one or the other; rather, they signal their recognition of—by conceding to, dismissing, or conspicuously ignoring—the contradictions that their notion of creative imitation presented.

Perhaps we can helpfully highlight the problems if we return to a comparison among texts. Against the background of Petrarch's and Erasmus's use of the digestive imagery and anticipating its later, more intricate use by

Bacon and Jonson, Sidney's position, both temporally and theoretically in the middle, seems curiously all-embracing. After he talks about the need for imitation of other authors, Sidney takes some time to talk about the improper practice of this type of imitation:

> Truly I could wish, if at least I might be so bold to wish in a thing beyond the reach of my capacity, the diligent imitators of *Tullie* and *Demosthenes* (most worthy to be imitated) did not so much keep *Nizolian* Paper-bookes of their figures and phrases, as by attentive translation (as it were) devoure them whole, and make them wholly theirs.
>
> (p. 202)

Acknowleding that his ideal proposal is one impossible to realize ("if at least I might be so bold to wish in a thing beyond the reach of my capacity"), Sidney offers a rather confusing rule for devouring texts and for "making one's own the external text in all its otherness." He first associates imitation with translation—a merger that Puttenham is careful to destroy ("not . . . as doth the translator"). Then—while seeming to contradict Bacon's subsequent distinction between books to be "chewed and digested" and those to be only "tasted" or "swallowed," and Jonson's advocacy of those with stomachs that "concoct" and "divide" over the creature who "swallowes, what it takes in, crude, raw, or indigested"—he proposes that imitators "devoure them whole." Yet his final phrase transforms this simply activity of indiscriminate ingestion into the more intricate activity of digestion, which assimilates, selects, converts, and nourishes: "and make them wholly theirs." In Sidney's *Apology*, the activity disparaged by other authors produces the results that they advocate: the external text is kept wholly intact, yet it is detached from its original moorings and becomes, through some process not explained, identified wholly as the product of its new owner.

Here the basic problem is evident: the two inherently contradictory procedures are brought together, accompanied by the writer's admission that this is, at best, a *desire,* "a thing beyond the reach of my capacity." To return to our beginnings, Petrarch's comments in a third letter hit a resounding note of unease and qualification that, as I have tried to show, echoes throughout the Renaissance. Invoking, as usual, the Senecan counsel that in imitating, writers should imitate the bees, he offers a broad disclaimer ("If after a trial you discover that it is ineffectual, you must blame Seneca"), as if anticipating the failure of anyone who tries to follow his proposal, which he describes—as elusively as he did in other places—as "an astonishing process."[29] The crux of the problem again is expressed in an anecdote:

> Macrobius in his *Saturnalia* reported not only the sense but the very words of Seneca so that to me at the very time he seemed to be following this advice in his reading and writing, he seemed to be disapproving of it by what he did. For he did not try to produce honey from the flowers culled from Seneca but instead produced them whole and in the very form in which he had found them on the stems. Although how can I say that something another wrote is not mine, when Epicurus' opinion, as recorded by Seneca himself, is that anything said well by anyone is our own?[30]

Even firm proponents of creative imitation inevitably betray it in practice. Furthermore, can we—or should we—make clear distinctions of possession and authorship between those who first produce texts and those who later adopt them? However, Petrarch does not follow through this line of questioning; instead, he repeats his view that, in any case, it is more desirable to imitate the bees, writing "in a style uniquely ours although gathered from a variety of sources"—and even better, "to produce [our] own thoughts and speech." Yet he concludes, nonetheless, that "in truth, this talent is given to none or to very few."[31]

In this letter, as in the others, the qualifications imposed on man's ability to work according to Seneca's advice follow—or are followed by—the introduction of some comments on the relation between the human artist and the divine artist. For Petrarch, this relation impresses more deeply the limits of the poet, for in this context, as in all others, "perfection is the property of him alone from whom we derive all our knowledge, all our possibilities,"[32] and "We must therefore be content with the limits of the talents that God and nature granted us."[33] When man is, in any way, considered creative, a comparison to the divine Creator, either implicit or explicit, becomes imperative. God becomes the ultimate touchstone to which the poet must be referred and against which he must be measured. This idea is also found in Puttenham, when he describes the poet as a follower who can "expresse the true and lively of every thing is set before him": suggesting, in other words, that the poet presents images—or imitations—of God's creations. But the comparison with divinity is also present in—and presents complications for—Puttenham's notion of poet as maker. As Colie notes when discussing his analysis:

> To call a poet a maker involves the question of category: if there is, in fact, but one Maker, the Creating God, then all other makers are themselves imitations, constructing as best they can imitations of God's creation. But if poets are really makers, then the category of maker must be enlarged to include them, and though there can be degrees of excellence within the category, the essence of all makers is the same, and thus, presumably, the essence of their creations is the same as well.[34]

The greater the emphasis placed on creating in "creative imitation"—the more the poet is called a maker—the greater the need to define his relation to God. For if man is truly a maker, then his activity can be no more circumscribed than God's (a concept that borders on blasphemy); yet if God is the only true maker, then the poet's work is more subservient and imitative. In the first half of the *Apology*, where images of freedom and creativity abound, the connection to heavenly creation is made quite often, and at the most crucial points, just when Sidney seems to be on the verge of describing human creation as divine. The poet does indeed "growe in effect another nature," "freely ranging onely within the Zodiack of his owne wit"—in essence, "making," as God does, with unrestricted license. Yet this section concludes with a sobering recognition (and evasion) of the implications of this analogy—that is, with an acknowledgment of a hierarchy:

> . . . give right honor to the heavenly Maker of that maker, who, having made man to his owne likenes, set him beyond and over all the workes of that second nature, which in nothing hee sheweth so much as in Poetrie, when with the force of a divine breath he bringeth things forth far surpassing her [Nature's] dooings . . .
>
> (p. 157)

In other words, as MacCaffrey has written, Sidney, along with other "sixteenth century poets, like their medieval counterparts, were reluctant to detach the poet's golden worlds from the actually of the divine Creation or to claim absoluteness for poetic creativity."[35] If a poet makes the move of independence away from both imitation of other authors and mimetic verisimilitude, then he confronts the crucial challenge of the divine Creation, which imposes and impresses the fact that the scope of human creativity is not absolute or self-sufficient, that poets will always be imitators, either of what Sidney calls the "inconceivable excellencies of God" (p. 158), of his creations, or of his creative activity itself. Yet, as these quotations suggest, there is a more liberating side to the issue. Sidney's honoring of the "heavenly Maker of that maker" allows him to conclude that the poet "with the force of a divine breath

. . . bringeth things forth far surpassing her [Nature's] dooings"—that the poet's function does indeed approximate, even mirror, God's.[36] If consideration of the divine imposes consideration of human limitations, it also reveals a close alliance between human and heavenly creation; the analogy between human and divine art suggests not only the God-given but also the God*like* power to make that humans have within their reach. God is the source of human creativity and the source of its limits, as well as the prototypical figure of limitless creation with which the poet can be associated, not only from which he must be differentiated.

Who, then, can claim the rights of authorship over any text—the model or the imitator, the source or the product, the subject or the writer? At what point can an author be said to have made someone else's words his own? At what point—and can we determine at what point—does the poet (as he almost surely will) stop imitating creatively and begin copying servilely? Be it God, or nature, or the works of other writers, or the subject matter itself that is considered as original source, where and when can the poet claim "the authoritie of his penne"? Furthermore, at what cost—to all sides—can any position be maintained? Inevitably, one part of the equation—the writer, his source, his text, or his subject—is eliminated when another emerges too forcefully. The balance between individual creations and an original design is uneasy, with one side of the scale always threatening to destroy the other. Marvell, standing at the end of the period I am discussing, presented the multiple potentials inherent in the relationship:

> This *Scene* again withdrawing brings
> A new and empty Face of things;
> A levell'd space, as smooth and plain,
> As Clothes for *Lilly* stretcht to stain.
> The World when first created sure

Was such a Table rase and pure.
Or rather such is the *Toril*
Ere the Bulls enter at Madril.[37]

When created Nature disappears as model, the landscape becomes both "new" and "empty," like the untouched painter's canvas (and like, I might add, the empty desert confronted by the dreamer in the *House of Fame*), ready to receive and become the artist's conception of the world. The artist's activity (and its scope) is at first likened to the act of original Creation, but the analogy to the divine artist—natural, life-giving, and free—is then restated (more precisely, it seems) and is transformed into an image of a ritualized and destructive human enterprise that endangers not only the object of the activity, but the human actor as well, who initiates, performs, and ultimately brings about the dissolution of the very process in which he is engaged.

II

Sixteenth-century love sonnets seem to bring into focus, either as foreground or background, all these converging concepts. On the one hand, the sonnet sequence is entrenched in tradition and convention, a "poetic form where novelty is both rare and often a hindrance . . . to expression";[38] on the other hand, the tradition of the sonnet itself "aimed above all else at the voicing of personal perceptions."[39] Modern commentaries on the sonnet sequence often take as enigmatic a form as do those on the issue of imitation that I have previously discussed. Maurice Evans, for example, speaks of the "prodigality of language and invention which the conventional nature of the sonnet themes encouraged by freeing the poet from the labour of creating his own situations."[40] Patrick Cruttwell notes that the quarrel between individuality and convention has "plagued the sonnet about love almost from its beginnings," for the sonnet sequence "early became . . . a conventional medium for the expression of Courtly Love . . . but at the same time always adopted the literary form of the autobiographical, the insistent 'I', and always claimed that 'my' experiences . . . were unique . . ."[41] Several critics have ascribed more authorial self-consciousness to this aspect of sonnet writing; William Nelson, for example, comments that "The ingenious poet . . . attempted rare and remarkable combinations, and the poet of power transcended the convention, but their election of the form showed that they willingly imposed upon themselves those limitations that they wished to challenge."[42] Perhaps the most eloquent statement is made in the seminal work of Theodore Spencer, who writes that

> In the sixteenth century, this saving loss of personality, this discovery of self through submission to an "other," could be accomplished to a considerable extent through convention. Convention is to the poet in an age of belief what the *persona* is to the poet in an age of bewilderment. By submission to either the poet acquires authority; he feels that he is speaking for, is representing, something more important than himself . . .
>
> But the submission to convention is by no means a passive process. . . . The convention . . . must obviously be freshened by continual re-examination so that it is re-made every time it is used.[43]

The convention frees; it is deliberately invoked so that it can be transcended; by submission to it a poet acquires authority because he "re-makes" it even as he uses it. These excerpts sound very much like the easy reciprocities that critics often read into and out of the Renaissance notion of creative imitation but that are rarely given such comfortable expression by sixteenth- and seventeenth-century writers. The sonnet sequence, for many reasons—only one of which is its weight of convention—almost guaranteed that similar issues would come into play and that the medium would provide a context for expressing the various problems those issues generate.

In the Renaissance, the relationship between love and poetry, writing and wooing, poems and passion, was strong and complex. Donne discovered that "To love . . . tribute of Verse belongs," and yet that "I am two fooles, I know, / For loving, and for saying so / In whining Poetry."[44] Love might deserve and demand poetry, but the association may or may not be a happy—or a possible—one. Raleigh decided, quite expediently, that only shallow passions speak, "but the Deep are Dumb"; witty eloquence cannot coexist with truly felt emotion: "They that are Rich in Words must needs discover / That they are Poore in that which makes a Lover."[45] The passion could either inspire or inhibit the act of writing; love poetry could excite rather than subdue the passion that could make writing difficult. And the secular poet did not have a monopoly on this relationship; Herbert dealt with the matter by concluding—or being forced to conclude—that (divine) love not only generates the proper poem but has already written it (*"There is in love a sweetnesse readie penn'd: / Copie out onely that, and save expense"* ["Jordan II"]). Spenser, like Donne, found that trying to write about love, instead of offering relief, aggravated the passions and the problems:

> But when in hand my tuneless harp I take,
> then doe I more augment my foes despight:
> and griefe renew, and passions doe awake
> to battaile fresh against my selfe to fight.
> (*Amoretti* 44)

Spenser also admits, in one sonnet, that his love prevents his writing *The Faerie Queene,* for he is plagued with a "troublous fit, / of a proud love, that doth my spirite spoyle" (*Amoretti* 33). Yet later he finds the relationship more rewarding: his "contemplation of [my love's] heavenly hew, / my spirit to an higher pitch will rayse"—in other words, to "sing my loves sweet praise" will prepare him to finish *The Faerie Queene*. But this inspiring interlude of love sonnets is a pastime, not a vocation, a period of "rest" that follows and then precedes a period of "toyle" (*Amoretti* 80). This linkage of love, leisure, and poetry was common. It was not only, as E. C. Pettet notes, that counter to the earlier stages of romance when love was proved by martial prowess, the "pen replaced the lance" and "poetry-writing came to be an indispensable part of courtship"; or that, as Robert Kellogg notes, there is an "explicit characterization of the speaker in the Renaissance sonnet sequence as both a lover and a poet."[46] As Neil Rudenstine suggests, "The enemies of poetry in the sixteenth century had a great deal in common with the enemies of leisure and love, and their lines of attack were likely to be similar"; consequently, Sidney, for example, was not atypical in that he "expected any apology for poetry to involve an apology for love."[47] It should not be surprising, then, to find the issues that concerned poets reflected in the debates about love. In fact, the relationship between lover and beloved often was discussed in terms that mirror those invoked to describe the relationship between a source or model and the work produced by creative imitation. Whereas, for instance, Petrarch argues against too close a resemblance between an original and the imitation of it, for "Certainly each of us has naturally something individual and his own in his utterance and language as in his face and gesture,"[48] and whereas Jonson, on the other hand, advocates that a writer should choose another writer and "so to follow him, till he grow very *Hee*: or, so like him, as the Copie may be mistaken for the Principall," we find Musidorus, in the *Old Arcadia*, arguing both for and against a similar type of relationship—and transformation—in love(rs):

> For, indeed, the true love hath that excellent nature in it, that it doth transform the very essence of the lover into the thing loved, uniting and, as it were, incorporating it with a secret and inward working. And herein do these kinds of love imitate the excellent; for, as the love of heaven makes one heavenly, the love of virtue, virtuous, so doth the love of the world make one become worldly. And this effeminate love of a woman doth so womanize a man that, if you yield to it, it will not only make you a famous Amazon, but a launder, a distaff-spinner, or whatsoever vile occupation their idle heads can imagine and their weak hands perform.[49]

Love imitates, according to Musidorus, and transforms the imitator into an image of the beloved—and imitated—object. Depending upon whether the love is divine or human/worldly, this transformation can be either "virtuous" or "vile." Pyrocles's answer recalls the idea that, ideally, writers should imitate in order to form their own distinctive style and texts:

> Neither doubt you, because I wear a woman's apparel, I will be the more womanish; since, I assure you, for all my apparel, there is nothing I desire more than fully to prove myself a man in the enterprise.[50]

Pyrocles considers the borrowed aspect of the woman as apparel that furnishes him with the means to achieve his true purpose: the development of his personal stature and identity as a man. But against Musidorus's *declaration* that the "very essence of the lover [is transformed] into the thing loved," which is the thing imitated, Pyrocles's response that the imitation will ultimately transcend the beloved object has the status of a "*desire*," not an affirmation. Substitute writing for loving in these passages, and the exchange echoes the debates and controversies surrounding the notion of creative imitation.

In the sonnets, the connection between writing and

loving is made quite automatically, and the relationship is often presented as inescapable and binding, not as willingly and willfully invoked or accepted. Drummond, for example, found not only that writing and loving were inextricably bound together but also that he was bound to engage in them:

> Know what I list, this all can not mee move,
> But that (*o mee!*) I both must write, and love.[51]

Fletcher presents a small history of his encounter with writing and loving: he wanted "A Poet to become," but "PHOEBUS denied"; then Venus came to him, and he accepted her invitation. But the bargain with the Goddess of Love is two-faced:

> That poison, Sweet, hath done me all this wrong;
> For now of Love must needs be all my Song.
>
> (*Licia* 1)[52]

Love begins as a substitute for poetry, one that is both enabling and restricting: now the poet can write, but he can only write of love. The woman herself is the source of his poetry; without her, he is no poet:

> But when your figure and your shape is gone;
> I speechless am, like as I was before.
>
> (*Licia* 47)

The beloved as the poet's inspiration is a common theme; the relationship is summed up in Shakespeare's comment about the subject of his verse: "thou art all my art" (Sonnet 78).[53] But the beloved who, muselike, supplies—and helps write—his poetry may also *prevent* his writing of poetry:

> O, blame me not if I no more can write!
> Look in your glass, and there appears a face

> That overgoes my blunt invention quite,
> Dulling my lines and doing me disgrace.
> (Shakespeare, Sonnet 103)

This source of the poet's inspiration is also a source of restrictions, a controlling, overwhelming, original design. The beloved may provide the words for his poem, but in doing so may make the poet a prisoner of the beloved's art; as Drummond suggests, his lady's words may seduce and overpower:

> And you her Words, Words no, but Golden Chaines
> Which did captive mine Eares, ensnare my Soule,
> Wise Image of her Minde, Minde that containes
> A Power all Power of Senses to controule.
> (Drummond, Sonnet 13)

Even as the seemingly benign muse or model, the lady, her love, and her art prove troublesome. At best, as a model, she (like love) is unstable, evasive, and fleeting:

> It is a vision seeming such as thou,
> That flies as fast as it assaults mine eyes;
> It is affection that doth reason miss;
> It is a shape of pleasure like to you,
> Which meets the eye, and seen on sudden dies.
> (Lodge, *Phillis* 26)

It is imperative, however, that the woman be thus intractable, that she insist on eluding the man even as she provides him his material and inspiration, that she remain incapable of being accommodated as both his lover and his artistic source. If the man cannot write without the woman as his model, neither can he write if she performs as an acquiescent and malleable support or complement. He is equally impotent, both as poet and lover, when her standards are completely adjusted to his and when her standards do not exist. Daniel, for example, questions whether he should continue to write when his verse meets with such unaccommodating resistance, but he concludes by realizing that without such resistance, he would not be able to write at all:

> Favours, I think, would sense quite overcome;
> And that makes happy lovers ever dumb.
> (Daniel, *Delia* 17)

The tension between the lover and his beloved, between the woman's art and the poet's, thus becomes the energizing force behind the poetry. It is the very unwillingness of the imitated object to acquiesce to the writer's mode that supplies him with his own sphere of activity. On the one hand, the lady's obstinate, challenging, overpowering, threatening stance subsumes his voice and restricts his freedom as a writer; on the other hand, if she were a manageable subject, if there were a possibility that she would agree to all his maneuvers, then he would be likewise restricted, for the intractability of the lady provides the poet with the freedom to make his most important move. If his poetry can exist only if it submits to hers, then the poet must surrender himself completely and identify all that he writes as his lady's. Daniel, in fact, in his dedicatory sonnet, *requests* that his lady and patroness accept this kind of transfer of "*my humble rhymes*": "*Vouchsafe now, to accept them as thine own! / Begotten by thy hand, and my desire.*"[54] This move, which costs the poet his title and vocation, can be made only because the lady will ultimately refuse the merger that his surrender will entail. Indeed, Daniel's initial offer to relinquish his poetry to his lady would involve her in a cooperative effort that implicates both of them as inseparable partners in the poetic enterprise:

> *Begotten by thy hand, and my desire;*
> *Wherein my zeal, and thy great might is shown.*
> *And seeing this unto the world is known;*
> *O leave not, still, to grace thy work in me!*

To accept the poet's gesture would be to compromise the lady's independent existence as both poet and person. As A. C. Hamilton notes, the woman in the sonnets "cannot call the lover's bluff. Hence, though he is bound by love, he remains free to yield himself entirely to her without fear of utter personal annihilation to which his words commit him."[55] Her resistance allows him to make the verbal surrender of his authorship that in the end allows him to continue to write. That complete loss of personal and artistic identity which seems to be always risked in both poetry and love (secular and divine) exists perpetually in abeyance: as ultimately inevitable but never fully consummated; as both fulfillment and demise; as an option that seductively beckons and then repels, that is courted and then denied. Like the model text that refuses to surrender its original moorings, the recalcitrant woman refuses to be assimilated by the poet/lover. Like the writer who imitates a prior work in an attempt to form a new, personal, and independent one, the lover must acknowledge his ties to and reliance on the woman, without losing his own identity as man and writer. The tensions between lover and beloved are the same as those between writer and model: in imitating a text or loving a woman so as to possess the object completely, the process may be reversed, and the writer/lover may himself become possessed, subservient to the other that insists as strongly on its own and independent identity.

III

For George Herbert, the proper subject and object of poetry (divine love) must be acknowledged as the final source and author of poetry; this recognition requires a painful and often resisted surrender (to God) of the poet's claim to the creative powers and independence of his art. The secular sonnet sequences of Sidney and Spenser demonstrate as well the challenge presented by their subject (human love). The beloved woman is portrayed as containing and exerting an art of her own that claims priority and superiority: as source and model, it inspires the poet, but it also threatens to subsume, deny, or, at the least, diminish his art. In this way the tension between creative autonomy and authoritative sanction inherent in the relationship between model and copy can serve as an analogue for the relationship between lover and beloved. The desired relationship with the lady seems to require that the speaker succumb to her poetic; what must be reconciled is his refusal to relinquish the possibility of his authorial autonomy and the need (poetically and personally) to acknowledge and conform to the standards presented by the lady. To "make no Thine and Mine" (to use Herbert's phrase from "Clasping of Hands") is the goal; the method is to make no thine and mine between the aesthetic principles that each advances, without totally dissolving the object of the speaker's poetry (the lady) or obliterating the identity of the speaker himself. The poets must perform that "astonishing process" on the women, and "by attentive translation (as it were) devoure them whole, and make them wholly theirs," incorporating but not destroying the "external otherness" of the woman and not renouncing that "something individual and his own in his utterance and language" or allowing "the very essence of the lover [to be transformed] into the thing loved."

Similarities between Herbert and Sidney have often been cited. Louis Martz, for example, notes poems in *The Temple* and *Astrophil and Stella* that share a concern for finding an effective mode of literary expression. Martz comments only on resemblances: their "search for simplicity," "their condemnation of elaborate modes of art," and "their insistence on the values of simple truth," all of which "sometimes leads to the presentation of the speaker as love's simpleton."[56] Other critics have suggested interpretations that distinguish the ostensibly similar techniques and solutions advocated by the two poets. Herbert's rejection of convention in favor of the plain style in "Jordan I" and the account of his attempt to find a suitable style in "Jordan II" have been compared most particularly to the first and third sonnets of *Astrophil and Stella*. Stanley Fish discusses the relationship between the final couplet of "Jordan II,"

> *There is in love a sweetnesse readie penn'd:*
> *Copie out onely that, and save expense,*

and Sidney's first sonnet, in which the poet is told by his muse to "looke in thy heart and write."[57] There is a "radical difference," Fish claims; Sidney is "advised to call on his own resources," while Herbert is "reminded that his resources are not his own."[58] Colie makes a similar point, that "where Sidney, excellent humanist that he was, had found sufficient truth in his own heart, Herbert took the divine love as his model."[59] This important and valuable distinction holds only for these poems, however; examination of a wider selection clarifies a significant characteristic of the two poetic stances. Sidney's third sonnet affords, in one sense, an even closer parallel to Herbert; there the poet concludes, after rejecting conventional and aureate language, that

> all my deed
> But Copying is, what in her Nature writes.

Here Sidney seems to be denying his artistic independence and the efficacy of his own words; like Herbert, he must copy what has already been written. For both poets, the model for imitation has shifted, but imitation remains the dominant aesthetic principle.

There is a crucial difference, however, not in the apparent meanings of the statements, but in the attitudes and situations of the speakers and in the processes that elicit these concluding remarks. In Sidney's poem, the final words that advocate copying are those of the poet, while in "Jordan II" another voice enters to impose this standard. When another voice directs the speaker of the sonnets, it tells him, essentially, to invent on his own;[60] it is the poet himself who (in Sonnet 3) proposes the aesthetic of copying.

Although the final lines of Sonnet 3 of *Astrophil and Stella* are analogous to the final lines of "Jordan II," then, the poem as a whole (and the authorial position) seems more closely related to "Jordan I."[61] In "Jordan II," as we have seen, the concept of copying is imposed by a voice that intrudes and usurps the poet's authorial status. In "Jordan I," however, the poet *chooses*, in a self-elevating poem that sets him apart from other, less admirable poets, to place certain restraints on his art. The poet bends to the demands of his subject, but he does so at his own will, proud of his stance, retaining credit for and control over his art. In Sidney's third sonnet, the speaker, with an equally clear sense of superiority, rejects the ridiculed literary techniques of "daintie wits" and "*Pindare's* Apes" who, even though they use "new found Tropes" and "strange similies," do so to deal with "problemes old" that provide the sanction for ("Ennobling") their witty inventions. The poet proudly claims, isolating himself from the masses and carefully making the distinction, that "For me in sooth, no Muse but one I know." Then, for a moment, the tone changes; the speaker admits a sense of regret. It seems that he does not simply *choose* to dispense with the style of other poets; instead that style is, for some reason, beyond his grasp. He frankly cannot *afford* to employ it:

> Phrases and Problemes from my reach do grow,
> And strange things cost too deare for my poore sprites.

This sense of the poet's lack of control prepares for the expressed surrender of the conclusion. But the suggestion, in the first line above, that the poet's restricted scope is the result of personal limitation, is transformed, in the second line, into a question of expenditure, and this maneuver allows the final lines of the poem to move back to the original stance of mastery and free choice. The solution hit upon ("How then? even thus:") is voluntarily (in his own voice) introduced and accepted as a sound financial venture. Unrestricted poetic range itself exacts a prohibitive price: the stylistic characteristics of other writers—both their ingenuity in devising "new" constructions and their aping of conventional modes—demand a "cost too deare" for this poet, and the principle that provides a more economical program for him is to copy what he reads in Stella's face. The speaker retains control over his poem; he can pronounce the resignation of his claim to autonomous authorship because he is, as the poem implies, confident that in the end this gesture is cheap and will entail no real losses: the investment will be less, and the rewards greater.

With this contract established, then, the poet can capitalize on his returns. In Sonnet 50, for example, in a direct address to Stella, whose name opens this poem, he explains that his thoughts of her "Cannot be staid" until they are expressed in words—words that portray *her* "figure." Yet "as soone" as he writes, he humbly continues, he can immediately discern the "weake proportion" of his craft as compared to the model it attempts "To portrait." There is no contest, according to the poet: he must "write my mind," but "thy figure" ("that which in this world is best") clearly is—and always will be—superior to him and his art. Beside her, his art is so diminished, falls so far from the mark, that it cannot even be allowed to survive. There is no choice, the despairing poet reports:

> . . . I cannot chuse but write my mind,
> And cannot chuse but put out what I write.

The poet's humility, his lack of resistance to the subservient position he accepts for himself and his art, seems absolute. He is very careful, however, to make it clear to his audience that, despite his complete acceptance, the admission is extremely painful for him; he views the situation "with sad eyes" and becomes pathetically sentimental about the fate of his poetic offspring ("those poore babes their death in birth do find"). But, regretful and resigned, the poet seems prepared to do what he must: to commit artistic suicide, to take his pen and cross out what he has written. At the moment of execution, however, the poet suddenly stops himself and uncovers a new factor—Stella's preserving grace:

> And now my pen these lines had dashed quite,
> But that they stopt his furie from the same,
> Because their forefront bare sweet *Stella's* name.

To deface his poem would be to deface Stella. She whose "figure" demanded the death of his poetry now demands its salvation. The relinquishment of his art and his own poetic identity has furnished him and his poem their inalienable right to life. Full of regret, the poet prepares for and makes his surrender to Stella, only, in the end, to reveal triumphantly that his surrender is his victory. Abdication of his role as poet allows him to occupy it all the more forcefully; his presentation of Stella as ultimate touchstone for his art provides the ultimate sanction for it as well. The shift to the third-person pronoun in the last lines (quoted above) makes clear that he has always been in control. It is not really Stella who stops his pen from blotting out the poem; the poem itself, the very lines that the poet has written, prevents the pen from murdering the words that have issued from it. The poet is all success now, not failure. He has paid a proper tribute to his lady; he has formally acknowledged and submitted to the eternal nature of her artistic superiority as his model, exposing the inferiority of his imitation, which does not, in comparison, even deserve to exist; and yet he does this all in a poem that, by the very principles he sets up to provoke his own demise, must be considered as inviolate as Stella herself. The investment has indeed been small: lip service is paid to Stella's inimitable qualities, but in the end, the poem becomes the unassailable object. What remains as savior of both Stella and the poem is the poet himself.

As Rosalie Colie suggests, Sidney makes a habit of "contradicting his expressed statement by the operation of the poem in which that statement occurs."[62] He can, as my two examples show, verbally renounce his poetry and in the process do everything possible to protect and validate it. For instance, as Colie notes, in Sonnet 90 the poet claims, "patently untruthfully," that

> In truth I sweare, I wish not there should be
> Graved in mine Epitaph a Poet's name.

The poem is written to convince Stella that he does not "by verse seeke fame," for she alone dominates every aspect of his life ("Who seeke, who hope, who love, who live but thee"). "Thy lips" are "my history," he explains, and the only praise he values is hers. Soon, however, he is protesting not simply that he does not seek fame but that the fame he admittedly does receive is borrowed, since "my plumes from others' wings I take." The exact terms of the debt are made clear in the concluding lines, which, looking forward to Herbert's remark in "Assurance" that "Thou didst at once they self indite, / And hold my hand, while I did write," acknowledge that

> . . . nothing from my wit or will doth flow,
> Since all my words thy beauty doth endite,
> And love doth hold my hand, and makes me write.

Here the poet totally denies his artistic independence, picturing himself as a passive amanuensis. Yet this complete humility provides him with the very fame he began by pretending to renounce: now his artful transformation allows him to accept it as a gift from his beloved.

In *Astrophil and Stella,* the poet's declarations that the subject and object of his poems are their ultimate source and author are thus belied by his sense (often implied by the same process that elicits these statements) of his own creative authorial power and his control over his subject. What for Herbert is often an uncomfortable and self-denying realization is employed by Sidney as a technique: although his declarations may be self-effacing in their expressed meaning, the poet loses nothing by speaking these words, which are chosen by rather than imposed upon him. He willingly and boldly assumes a posture that surrenders his art to Stella—an effective stance not only for the successful poet but also for the persuasive and seductive lover. For Stella is flattered while the poet accrues for himself his poetic identity, and, in all these poems, the poet constructs an arrangement that makes their activities mutual: her fame becomes his, her inviolability becomes his, her art becomes his. To borrow a phrase from Herbert again, his poetry thus becomes "that which while I use / I am with thee, and *most take all*" ("The Quidditie"). Rosemond Tuve is correct to feel that Martz's analysis of resemblances between Herbert and Sidney is one-sided and inadequate; in reference to "Jordan I" and the first and third sonnets of *Astrophil and Stella,* she claims that the themes of the two poets "are a world apart": Herbert is primarily concerned not with literary style but with "living a life" infused and informed by divine love.[63]

Although Tuve's point about Herbert is well taken, her analysis seems inadequate as well; it is not clear that Sidney's only concern was with the stylistic problems of writing a poem. Many have noted how Sidney—through the dramatized Astrophil—assumes different positions in different poems; each is an instrument employed to satisfy the rhetorical purpose of uniting speaker and beloved.[64] When the poet renounces the creative independence of his art, he chooses to do so for its effect on his relation to Stella; he uses this as a ploy to dissolve the distance between himself and his lady. As Richard Young suggests, "Sidney has exploited the technical problem, the relation of manner and matter, as the chief means of presenting the dramatic problem, the relation of lady and lover."[65] When, in Sonnet 3 or 90, for example, Stella is presented as the author's artistic source, when her art becomes his, then not only do the model and copy become identified, but the separation between lover and beloved is resolved as well. Sidney takes this stand for its rhetorical and dramatic effectiveness; he makes such gestures as part of a strategic role that he himself produces and willingly plays. The poet who states "I am not I, pitie the tale of me" (Sonnet 45) is the poet who is willing, in triumph, to pretend to surrender his individual authorship as he self-consciously turns himself into the main character of his own tale—a poet willing to deny himself so that he may create an effective artifact.[66]

Spenser never asserts to the same (extreme) degree the power of his beloved over his poetry. When the "external otherness" of the lady and her art threatens the poet's artistic independence (and existence) in the *Amoretti*, the poet tries to negotiate between the pressure to succumb to her power and his reluctance to deny the possibility of his own creative achievements. Whereas Sidney may provide for himself a context in which expressing the surrender of his claim to authorship does not cost him anything, Spenser consistently (and characteristically) accommodates both sides of the battle for artistic superiority and control: the victorious poetic source is both himself and his lady, both his creative imitation and her imitated creation.[67]

Modern criticism has tended to emphasize the lack of felt tension and the avoidance of intense conflict in the *Amoretti*. Hallett Smith, for example, claims that for Spenser, in contrast to Sidney, "Love is not primarily a dramatic conflict."[68] William Van O'Connor, employing categories of "poetry of exploration" (which confronts and works through intellectual and emotional problems) and "poetry of exposition" (which unequivocally represents and illustrates an attitude), places Spenser's sonnets in the latter division.[69] J. W. Lever speaks of how in some of the *Amoretti*, Spenser transferred his technique for treating love in "allegorical romance," where "there was no need . . . to imitate the stresses of real life," into the distinctly different sonnet form, creating an "unperturbed, harmonious flow of exposition," and he explains in general that the formal technique of Spenser's sonnets sets up "fourteen lines of unhalting, melodious exposition."[70] Although these views helpfully highlight Spenser's retreat from accentuating dramatic conflict and explicitly engaging tension in favor of a more graceful, even effortless, style and tone, they seem misdirected in their claims that he avoids the conflicts, stresses, and tensions altogether.[71] Part of Spenser's technique involves the implicit acknowledgment and presentation of conflict that is alleviated (but not resolved) by the poet's ability to maintain two positions at once, to circumvent issues at the same time that he confronts them. The stresses are there; the process of the poems is designed by the poet's subtly and deftly (and often playfully) managed maneuvers that allow him to satisfy two conflicting demands. He advocates or succumbs to contradictory alternatives almost simultaneously, retreating from tension not by ignoring it (or opting for a single position) but by embracing it totally.

The conflict produced by the lady's challenge to the poet's art is encountered frequently in the *Amoretti*. Yet, although Spenser often directly confronts the problem, he ultimately avoids the necessity of resolution, of taking one stand and rejecting another. In Sonnet 17, for example, the poet seems to acknowledge the lady's superiority and power and to locate in her an original model that overwhelms any attempt to imitate or represent it: though he "devize at will," the poet cannot "expresse her fill." The implication is that she both defines and controls the artistic domain, and is thereby responsible for his failure to write well. Yet the accusation or complaint (which is also, of course, a compliment), before it is completed, turns into a confession of the poet's own intrinsic faults. His skill, he admits, is inadequate by itself, even when it is not being measured against the lady's artistic standard, even when untouched and undisturbed by her overpowering influence:

> The glorious pourtraict of that Angels face,
> Made to amaze weake mens confused skil:
> and this worlds worthlesse glory to embase . . .

The power of the lady is characterized as an art (the "pourtraict") "made" (with implied deliberateness) to threaten that of other men. But in the second line we find that the lady's art amazes or confuses not the poet's valuable and effective skill but rather the already "confused skil" of "weake" men. In the third line the retreat from accusation and from recognition of the lady's (destructively overwhelming) control is made stronger: the troubled condition of the world ("this worlds worthlesse glory") exists even syntactically prior to the lady's debasement of it. Before the poet fully acknowledges the lady's ability to distort and diminish his art, he entertains the evidence of his own failings. He resists taking either stand and manages to conflate his own power over his art with that of his lady. The possibility that the poet's skill is confused and weak without the lady's interference may imply the failure of his poetry, but it is a failure for which the poet may claim responsibility. At the same time, however, the poet bows to the influence and effect of his lady.

In his attempt to assert his own artistic identity at the same time that he acknowledges the control wielded by the lady, whose standards he wants to both follow and transform, Spenser often manages to merge their artistic activities. The issue of poetic independence and superiority thus becomes an instrument for harmonizing or identifying the poet with the beloved. In *Astrophil and Stella*, the speaker's surrender of his claim to authorship becomes a strategy for closing the distance between himself and Stella; in the *Amoretti*, the poet employs his more equivocating stance for a similar purpose. In Sonnet 23, the response to the lady's challenging (and consuming) art becomes a method for creating an alliance between speaker and beloved. The poet begins by describing the action of Penelope, whose art worked against her suitors but *for* Ulysses:

> *Penelope* for her *Ulisses* sake,
> Deviz'd a Web her wooers to deceave:
> in which the worke that she all day did make
> the same at night she did againe unreave.

Here Penelope's activity is clearly an artistic one ("Deviz'd a Web") that is both threatening (to the wooers) and supportive (for Ulysses), as well as creative (weaving the web) and destructive (unraveling). The poet has yet to make the connection to his own situation; he seems at this point to have two alternative positions in which to place himself: the role of either Ulysses or a wooer. In the next two lines, he carefully makes the analogy: his lover is Penelope, and he is one of the unfortunate suitors subverted by her art, her cunningly contrived and ominous "subtile craft":

> Such subtile craft my Damzell doth conceave,
> th' importune suit of my desire to shonne.

But in the next lines the roles change, and the conflict between the lady-artist and her suitor is presented as a conflict between two artist-lovers:

> for all that I in many dayes doo weave,
> in one short houre I find by her undonne.

Here the roles become subtly intermixed; the speaker, who weaves, assimilates the function of Penelope and his lady. His control over his art, however, is not as extensive as that of the two female artists over theirs; the poet retains the creative function for himself but attributes the destructive action (against his own art) of unraveling to his lady. The next lines serve to identify, differentiate, and confuse the roles even further:

> So when I thinke to end that I begonne,
> I must begin and never bring to end.

Here the speaker more fully assumes the function and identity of his lover through the unceasing repetition and renewal of his artistic activity. For a moment the poet allows himself to be Penelope, the subtle artist controlling his own fate and that of others. But the poet plans on completing his work, and in the next lines he re-presents the power to undo his art as his lady's:

> for with one looke she spils that long I sponne,
> and with one word my whole years work doth rend.

The poet has now managed to merge inextricably his activity with that of his lady. She no longer unweaves her own art; rather, she works upon the art of the poet, which he constantly begins as she prevents its completion. To fill Penelope's postion now requires the simultaneous participation of both the speaker and the lady, who are involved in a mutual, reciprocal occupation, however treacherous (for either of them) it may be. The final lines capitalize on the ambiguity and confusion of roles and reveal the speaker's strategy:

> Such labour like the Spyders web I fynd,
> whose fruitlesse worke is broken with least wynd.

"Such labour" has an ambiguous reference; it can refer to the activity of either the lady or the poet. The comparison to the spider does not help to clarify the matter; it is unclear here whether it is the poet or the lady who simulates the spider's activity. Furthermore, the spider's web has a double-edged meaning: it implies a capturing, deceptive, and subtly woven design as well as a tenuous and fragile piece of work. At this point, either (or both) characteristics may be ascribed to the lady's weaving and unraveling of her own art (to subvert the suitors), the lady's undoing of the poet's attempts to perform his artistic function, or the poet's own artistic endeavors when confronted with the lady's opposing action. The final line restricts the possible interpretations: the "labour" is described as "fruitlesse worke" that is easily "broken." Although the quality of the work is now defined (as flimsy and insubstantial), the worker is not. The conclusion may thus imply the futility of the lady's attempt to subvert the speaker's art, or the futility of the speaker's attempt to sustain and complete it. The poet carefully avoids the creation of a position that explicitly either asserts his artistic superiority over his lady or succumbs to her control over him and his art.[72]

Though not explicitly, however, these lines do—quietly—proclaim the poet's victory. The work described is "broken," but no agent of this destructive action is identified. In addition, the poem makes it clear that the destructive process of undoing the artist's work is equally renewing; if artistic endeavors are disrupted, they are disrupted only so that they may be revived and resumed. The relationship between Ulysses and Penelope was preserved by the repetitive creation and destruction of her carefully woven material; the success of her design depended on the dissolution and sacrifice of her art. Similarly, the poet preserves a relationship between himself and his lady by characterizing his art as constantly begun and undone. The poet's portrayal of the frustration of his artistic endeavors is a strategic maneuver that mediates between and incorporates him and his "Damzell" by synthesizing their artistic performances. His "fruitlesse worke" functions to engage and entrap the lady; indeed, while it is similar to the flimsy spider's web in that it is so often broken, it can also perform the spiderlike activity of subtly entangling and ensnaring precisely *because* it is so often broken. Thus here the two implications of "the Spyders web" are sustained; in fact, each meaning depends upon rather than excludes the other.

In Sonnet 23, then, the poet's ascription of an artistic activity to his lady at the same time that he presents himself as artist creates a reciprocity of action between them. His acknowledgment of and complaint against his lady's artistic power and his own subordinate position are also his instruments of triumph. The "subtile craft," the spiderlike weaving, always undone, and for that reason entrapping, is first attributed to the lady, but it is also the poet's: he devises a method for admitting his lady's control and skill while he asserts and presents the prowess of his own art—at once conceding that she overwhelms his artistic endeavors and illustrating that he can overwhelm her challenging art, granting her her resistance and distance while ensnaring her (in his poem) into a position that integrates their actions.[73]

Spenser takes a further step in Sonnet 28, which proceeds from the situation established in Sonnet 23, opening with the poet's anticipation that his art and that of his lady can—and will—be equalized and harmonized. The poem introduces the two in similar positions as artists:

> The laurell leafe, which you this day doe weare,
> gives me great hope of your relenting mynd:
> for since it is the badg which I doe beare,
> ye bearing it doe seeme to me inclind.

The laurel leaf, the poet's crown, is the possession of both the lady and the speaker. Here the speaker does not see the lady's art as threatening or frustrating; rather, he now presents their shared artistic status as a source of encouragement. The sense of harmony suggested here, however, is somewhat qualified by the laurel's symbolic connection with pride.[74] The implications of this second meaning are enlarged in the next quatrain, where the poet clarifies the relationship. The equality is tampered with and now the poet suggests the possibility of his own ascendancy:

> The powre thereof, which ofte in me I find,
> let it lykewise your gentle brest inspire
> with sweet infusion, and put you in mind
> of that proud mayd, whom now those leaves attyre.

Here "the powre" of the poetic laurel is first attributed to the speaker, who expresses his generous hope that his lady will be similarly endowed; in fact, he almost seems to be giving his permission ("let it . . .") for it to be shared with the lady. This is also an appeal or entreaty, however, implying that the lady ultimately controls the poetic realm and determines the extent of her own art. Again the speaker manages to allow for the superior position of both himself and his lady. The explicit use of the word *proud* intensifies the complication of the meaning of the laurel: pride is now associated with the lady as well as with the speaker, who claims original possession of the laurel and the power to infuse her with its essence.

The third quatrain completes the shift from a picture of reciprocal, mutually encouraging behavior to one of a threatening and dangerous struggle. The speaker takes the story of Daphne and shapes it to fit his own purpose:

> Proud *Daphne* scorning Phæbus lovely fyre,
> on the Thessalian shore from him did flee:
> for which the gods in theyr revengefull yre
> did her transforme into a laurell tree.

The intitial connection of the beloved to "Proud *Daphne*" defines her laurel as one of pride, but the reference to Phoebus maintains the poetic implications. In this version of the story, Daphne's flight from the dominion of the god of poets is presented as a threat, "scorning" or denigrating the power of the artist. But the speaker's conclusion of the story reverses the positions. Daphne's metamorphosis is not accomplished in answer to her own request or out of compassion or respect for her situation; rather, it is performed out of revenge and indicates her subordinate role. Her challenge to the power of the poet is converted into a threat directed *at her*; the transforming power itself does not lie with the lady. Yet the metamorphosis works in more than one way: while it places the lady in a manipulated position that she does not control, it also means, necessarily, that she bear the symbolic laurel, thus sustaining the original concepts of both her acknowledged poetic activity and the mutual identity of poet and lady. The final couplet thus reads as both a warning and a plea. A distance is admitted that still must be closed; the poet commands, threatens, and begs the lady to fill the role (of acquiescent and complementary poet and lover) he has created for her:

> Then fly no more fayre love from Phebus chace,
> but in your brest his leafe and love embrace.

In Sonnet 28, then, the poet begins by assuming a potentially harmonious relationship between his lady and himself, but, unable to sustain the image, he carefully acknowledges both the lady's threat to him and his threat to her. In Sonnet 29, the challenge of the lady's power to thwart and frustrate his art is renewed and made explicit, and the poet seems initially to concede her superiority. He opens with an invitation to the reader to examine how his art is disparaged, distorted, and transformed during its exchange with his lady, as his intentions are subsumed by hers:

> See how the stubborne damzell doth deprave
> my simple meaning with disdaynfull scorne:
> and by the bay which I unto her gave,
> accoumpts my selfe her captive quite forlorne.

The reader is asked to inspect the evidence ("See how . . ."), but the issues become extremely complicated as the poet attempts simultaneously to uphold the lady's standards and to maintain and proclaim the value of his own art. He claims that the lady perverts his "simple meaning": he has given her "the bay"—symbol of the poet—and she has turned it into the symbol of the conqueror. The lady, then, is charged with initiating the competition both by distorting the simplicity of his gesture as a poet and by transforming the meaning of that gesture into an indication of her prowess and superiority. She has even subverted the method by which the poet created a sense of his own power in Sonnet 28. There the poet implied that he was in the (superior) position of possessing artistic skills that could be transferred to her; now the lady has deprived him of that stance by contending that such gifts are offered by men who are—or who immediately become—subordinates.

Next the poet offers to the reader the evidence to support his claim; he ostensibly quotes the lady's own words that justify the complaint he has issued in the first four lines:

> The bay (quoth she) is of the victours borne,
> yielded them by the vanquisht as theyr meeds,
> and they therewith doe poetes heads adorne,
> to sing the glory of their famous deedes.

The lady's first two lines correspond to the poet's previous paraphrase; she claims that the bay is given as a reward to the "victours" from the "vanquisht." But the lady has said more, and we see that the poet's *too* "simple" restatement does not sufficiently describe her more complex response. She adds that the victors then in turn bestow the bay upon the poets who "sing the glory of their famous deedes." This returns to the bay the original (poetic) value the speaker had intended for it, and also returns it to the poet, thereby affixing a tone of reciprocity to the spirit of competition evoked by the lady's first two lines. If, in Sonnet 28, the poet could not sustain the sense of equality and harmony between him and his lady, here we find that neither can he sustain the sense of her controlling, distorting, and overpowering influence. Indeed, while he claims to be directly quoting the lady, the words he attributes to her open up the possibility of his succumbing to her challenge at the same time that he makes *her* words conform to his own purposes.

The transition back to the poet's own voice is subtle; there is no clear differentiation between the lady's manner of speaking and his own. The next line again makes explicit the sense of challenge and contest ("But sith she will the conquest challeng needs"), but the poet undermines the lady's "conquest" by making it appear that she has won only by his permission ("*let* her accept me as her faithfull thrall"). However, this line (like the similar one in Sonnet 28) also reads as a plea and indicates at the same time that even the poet's submission depends on the lady's acceptance. The poet's meaning is no longer "simple" as he attempts to reconcile his poetic endeavors with his lady's will, to assert his own victory and superiority while he formally accepts hers.

The concluding four lines of the poem completely destroy the simplicity the poet claimed for himself in his opening statement:

> that her great triumph which my skill exceeds,
> I may in trump of fame blaze over all.
> Then would I decke her head with glorious bayes,
> and fill the world with her victorious prayse.

The first of these lines may imply both that the lady's "great triumph" exceeds the poet's skill and that the poet's skill exceeds her great triumph. The next implies that the "trump of fame" will be sounding for the poet when he sings of his lady, as well as that he will use the "trump of fame" to proclaim *her* triumph "over all," including himself. The following line ostensibly returns to the poet's original gesture of offering the poetic bay to his lady, but the very wording of the description of the action indicates how the poet has modified his art in response to the lady's challenge: the simplicity of his earlier phrase—"the bay which I unto her gave" (line 3)—contrasts with the elaborate flourish of this later formulation—"Then would I decke her head with glorious bayes" (line 13). Furthermore, the preceding twelve lines of the poem prevent any simple reading of this line: the poet may be reoffering a poetic bay, or he may be assuming his possession of the conqueror's bay; in addition, the reciprocity suggested in the second quatrain implies that even by relinquishing the bays to his lady, the poet, by the lady's own admission and conditions, can expect to have the bays given back to him. The final line deftly claims victory for both the poet and the lady: "her victorious prayse" may refer to the poet's praise of the victorious lady, the poet's victorious praise of his lady (i.e., the victory of his art), and the lady's praise of the poet. The poet, then, has restored the possibility of his own artistic achievements while conforming to the strictures set up by the lady, and he demonstrates the power of his poetry while exhibiting the lady's capacity to modify, transform, and overwhelm his art. The reader "sees" the poet's ability to maintain and sustain his own poetic endeavors as well as the influence of his beloved on those endeavors, for the poem denies its original claim to simplicity in the process of asserting itself as capable of meeting the challenge of the lady.

Louis Martz has described the attitude of the speaker in the *Amoretti* as one of maturity and discretion, and he insists that "There is no danger that this discreet lover will ever lose his strong sense of duty and propriety."[75] Although Martz applies this observation to other aspects of the *Amoretti*, he correctly senses the stance of the poet who rarely leaves any traces of bald reproach or self-assertion, of rebellion or triumphant ascendancy. When the poet challenges the ability of his lady to form—or transform—his art, he simultaneously accepts it. There is a type of decorum in his engagement with this problem, whereby he subtly maintains the power and possibility of his own artistic achievements at the same time that he acknowledges the superior position of the lady and her control over his art, never unequivocally and unambiguously assuming a single position. Just as love may demand the surrender of personal independence and individuality, turning the lover into a servile image of the beloved, so may poetry demand the surrender of artistic authority and authorship, turning the writer into a servile copier of the model he attempts both to incorporate and to transform. While Herbert painfully and with frequent resistance submits to this demand, and Sidney boldly pretends to, Spenser accommodates it without compromising himself.

IV

As we have seen throughout this study, a poet's attempt to come to terms with an external model of authority requires that he also come to terms with the issue of poetic silence. For Chaucer, both the absence and the presence of an ultimate authority can completely overwhelm the individual voice, and yet the poet must silence *himself* to preserve the image of such an authority. In *The Faerie Queene*, the poet's resurrection of his own voice after the silence that follows Nature's decree ensures the possibility that his poem can be continued, yet he also desires to relinquish his voice because it destroys the authoritative image he so desperately seeks. And Herbert constantly discovers that he must give up all pretensions to writing poetry in his own voice, for "we must confesse that nothing is our own" ("The Holdfast"). It is God who "put the penne alone into his hand" ("Providence"); the essence of verse derives from and partakes of the creative power of divine, not human, art. Given the similar nature of the issues Sidney and Spenser engage in their sonnets—artistic control, creative autonomy, authoritative sanction—and the ways they choose to engage them, each must also at some point confront what is considered either the choice or the imposition of silence: the disappearance of the personal poetic voice when faced with a challenging claim to authorship and authority. The situations are somewhat different, since in the sonnets Sidney and Spenser are writing more insular poetry and can be more manipulative and even playful with their subjects. Sidney proposes a delightful solution to the dilemma of poetic silence—a dilemma he himself has created by presenting the object of his poetry as its author. Richard Lanham has suggested that Sidney's desire to obtain "grace" from Stella is a desire for a more physical than spiritual consummation of his love.[76] As he conflates the stylistic problem of merging manner and matter with the dramatic problem of merging poet and lover, Sidney finds a single solution for both issues: the kiss.[77] According to Castiglione, the kiss could serve such multiple functions, because the mouth is associated not only with the physical body but also with language. Whereas for the sensual lover the kiss is primarily a union of the bodies,

> . . . the reasonable lover woteth well, that although the mouth be a parcell of the bodie, yet it is an issue for the wordes, that be the interpreters of the soule, and for the inwarde breath, which is also called the soule.[78]

The act of kissing is in this way connected with the act of expression. Thus Sidney can discover that it is from Stella's kiss that the poet receives the ability to write his poems of love. In Sonnet 74, he first describes his confusion about why he can write so well despite having abandoned the conventions of other poets:

> How falles it then, that with so smooth an ease
> My thoughts I speake, and what I speake doth flow
> In verse, and that my verse best wits doth please?

He then prepares for his conclusion, creating theatrical suspense by rejecting unnamed possibilities:

> Guesse we the cause: 'What, is it thus?' Fie no:
> 'Or so?' Much lesse: 'How then?'

Finally he pulls the magic answer out of his bag of tricks:

> . . . Sure thus it is:
> My lips are sweet, inspired with *Stella's* kisse.

But this is only a partial solution, because if Stella is the source of his art, if it is through her lips that the poet can "speake . . . In verse," then he must silence his own speaking voice. Indeed, as he writes in Sonnet 81, he "faine would . . . paint thee to all men's eyes," but "she forbids." Hence, another crucial roadblock: the poet "cannot silent be," but the lady insists on his silence. The solution? Another expedient bargain:

> Then since (deare life) you faine would have me peace,
> And I, mad with delight, want wit to cease,
> Stop you my mouth with still still kissing me.

Her kiss provides him with his mode of expression *and*—without any contradiction, given the terms of his argument—also enables him to be silent. The possibility of writing poetry both begins and ends, for Chaucer, with the appearance of the "man of gret auctorite," and for the Spenser of *The Faerie Queene*, with the "Sabaoths Sight," and for Herbert, with the acknowledgment that "God is all"; Sidney, however, happily finds that both expression and silence can be achieved with a kiss from Stella.

Spenser, in the *Amoretti*, explains the problem in his own terms when he asks himself,

> Shall I then silent be or shall I speake?
> (*Amoretti* 43)

The strictures imposed on his art by his lady—the ultimate object and judge of his verse—"tie" his "toung" and demand that he not speak, but without his powers of articulation he would "in silence die." Not only would *he* die, however, and with him his poetry, but the delicate balance he has carefully constructed in his verse between himself and his lady would also perish. His answer is to fashion a new poetic style that will fulfill both of their needs. First he will teach himself to speak in silence:

> . . . I my heart with silence secretly
> will teach to speak . . .

And then—although he had initially named the lady as *his* artistic mentor and model ("Thus doth she traine and teach me with her lookes, / such art of eyes I never read in bookes" [*Amoretti* 21])—he will teach *her* how to read this silent speech:

> Which her deep wit, that true harts thought can spel,
> wil soone conceive, and learne to construe well.

He will, in short, with characteristic accommodation, devise a new mode of expression that will allow him to speak while conforming to his lady's requirement that he be quiet. Silence—the surrender of claims to personal authorship and to the authority of the individual voice—also strategically becomes a speechless assertion of aesthetic control and success.

That Spenser is able to equivocate so deftly and eloquently that neither he nor his lady must lose his or her separate identity or individual will is a testimony to the success of his artistic achievement: his ability to construct in a poem the conditions that bring together two conflicting claims to authorial superiority without destroying either one. Yet he is quite unlike Sidney, who openly admits that he creates the means for and terms of *his* success in a fiction, delighting as he unveils his well-crafted solutions. For Spenser, the notion that the balance he achieves is sustained only through his own writing is more disturbing, particularly when he sees through his own literary constructs to glimpse the antagonistic forces outside of them that always threaten their fragile structure. Several critics have mentioned the unsettling ending of the *Amoretti*, the oddly minor note of the last several sonnets, which seem out of place in this sonnet sequence and somewhat incongruous given the happy ending that we know about from biographical information and from the *Epithalamion*.[79] At the end of the sequence, an unanticipated challenge is received from a "Venemous toung" that intrudes upon the poet's enclosed world and creates "breaches" "in my sweet peace" (*Amoretti* 86). It is the tongue of slander, much like the Blatant Beast of Book VI of *The Faerie Queene,*[80] that disrupts the poet's vision and causes the separation between him and his beloved that is not resolved in the sequence. That the resolution is eventually achieved is revealed in the *Epithalamion,* but this poem clearly reflects its author's sense of the poem as a poem—as a fiction ultimately defenseless against the external world of destructive elements. "Wylde wolves" (69), "perrill and foule horror" (322), "false treason" (323), and "tempestuous storms or sad afray" (327) are all acknowledged as real conditions that the poet must continually ward off to sustain what is now seen as a very tenuous "sweet peace." The poet knows now that his private vision will eventually collapse under the onslaught of these external forces, and his request is a small one: "But let this day let this one day be myne" (125). The word *one* added to the repeated phrase in this line emphasizes the circumscribed scope of the poet's present aspiration.[81] He knows that in a short time his one day will be over, and it is up to him to make the most of it. So he uses his only weapon, his pen, to obtain his temporary peace and to create and preserve, for as long as possible, his private vision. For the *Epithalamion* is largely a series of evasions and delays that not only describes the wedding day (and night) but also protracts it. Every detail is discussed, every available accessory feature is brought in,[82] every possible digression is made; we view each aspect of this day as if in slow motion. The poet begins "Early before the worlds light giving lampe / His golden beame upon the hils doth spred" (19–20). It takes five stanzas of bidding his beloved to awake before she finally rises; we are made to view and admire her (along with numerous glances aside) as she comes from her chamber, as she enters the temple gates, as she stands before the alter. Then "al is done" (242), but not really; the bride must be brought home again, and only then do the celebrations begin. The poet anticipates the end of the day and the end of his beautiful vision, and he commands that it be recorded before the degeneration begins:

> . . . doe ye write it downe,
> That ye for ever it remember may.
> This day the sunne is in his chiefest hight,
> With Barnaby the bright,
> From whence declining daily by degrees
> He somewhat loseth of his heat and light,
> When once the Crab behind his back he sees.
> (263–69)

Although he has chosen "the longest day in all the yeare" (271), he knows that "never day so long, but late would passe" (273). As the "long weary day" draws to a close, however, he begins to regret its length because of the consequent brevity of the remaining night. But the poet persists: he turns the end of day into the beginning of night, which (complete with welcomes, preparations, prayers, evasions, and more diversions) lasts, given the twenty-four-hour/stanza design, for no fewer than eight stanzas. What the poet has done, essentially, is to expand his poem, within the temporal limits of its subject, as far as it can go,[83] and in doing so he prolongs his special day. Chapman's line in *Hero and Leander* could serve as this poet's motto: "I use digressions thus t'encrease the day."[84] For this one day is his only while his poem sustains it, and it will last as long as he can sustain the poem, which is his only defense—though a temporary one—against the inevitable decline. At his back is always the awareness that "it will soone be day" (369), the next day, and that his bliss and his poem will have to end. The resolution that the poem represents is of his own making and hence must sooner or later succumb to the external realities it sets itself up against. His song cannot endure forever ("cutting off through hasty accidents, / Ye would not stay your dew time to expect"); at best, it can be an "endlesse moniment" only "for short time."

This sense of the tentative and fleeting quality of the poetic resolution—and of the tentative and fleeting authority of the authorial voice—characterizes all the poems I have discussed. Sidney may have Astrophil find a most satisfying answer to the problem of keeping silent and writing, inseparably binding himself to Stella no matter what the demands of the moment, but his success is no more durable than Spenser's: by the end of his sequence that "Sweet kisse" of Sonnet 79 is just a memory. "*Stella* hath refused me" (Ninth Song), reports the bitter and grieving poet, who has been, against his will, "forst from *Stella*" (Sonnet 87) and must contend with "traytour absence" (Sonnet 88). He feels more forcefully than ever the "poysonous care" of "Envious wits" (Sonnet 104); he admits "rude dispaire" as his "daily unbidden guest" who intrudes "soone as thought of thee breeds my delight" (Sonnet 108), for, despite all the poet's labors, the undeniable and final reality is that "*Stella* is not here" (Sonnet 106). Similarly, no resolution in *The Faerie Queene* is more than temporary; all conclusive occasions eventually recede. Although the poet's voice and allegorical mode of expression can produce moments of coherence, they also render impossible—even destroy—any sustained final vision: Marinell and Florimell can be brought together, but not as harmoniously or completely as the Thames and the Medway; the Blatant Beast can be caught, but it cannot be restrained forever; Nature's decree can be reported, but it cannot be made to endure. In the *House of Fame*, the silence that interrupts and "concludes" the otherwise unending search for authority suggests the poet's need for an external authority to support, sustain, and sanction his poetry; it implies as well that this authority both requires and is bestowed by the curtailment of the poet's own voice and that such an authority may prove inadequate if allowed to function. Whatever the techniques—simple or elaborate—by which a poet maintains his authorial and authorizing privileges as creator and maker, he inevitably comes up against his own limitations and against other voices, models, or systems that claim priority or impinge upon his self-proclaimed area of autonomy. Often these models are sought, when the poet recognizes the insufficient authority of his own voice; often they intrude themselves into the space the poet has attempted to establish as his own. Yet the inadequacy or ultimate inaccessibility of such external sources and their inability to accommodate fully the poet's individual voice pressure him continually to come forth to be tested as his own principle of authority and his own source of sanction—even though he, too, is continually found to be lacking. The dream visions, *The Faerie Queene*, the *Mutability Cantos, Astrophil and Stella*, and the *Amoretti* all, finally, seem inconclusive or incomplete. The sense of ending is always sober and irresolute. It is the sense that the word is fragile; that the maker's claims to autonomy are, in the end, insubstantial; that his reliance on other voices will prove misplaced; that his individual creations cannot endure, either by assimilating or by resisting the influence of the world outside them; that a writer's authorship and the authority of his text are never final, never unassailable, never absolute.

Notes

Introduction

[1] See Edward Said's discussion of the "enabling" conditions of an author's authority and the inevitably accompanying restraints or "molestations" in *Beginnings: Intention and Method* (New York: Basic, 1975), esp. pp. 81–100.

[2] In other words, I do not here focus specifically on *political* or *social* forms of authority, although I would not want to deny the dangers of ignoring such contexts. For some recent studies of Renaissance literature—that complement this one—more oriented towards those contexts, see Stephen Greenblatt, *Renaissance Self-Fashioning: From More to Shakespeare* (Chicago: Univ. of Chicago Press, 1980) and his article "Invisible Bullets: Renaissance Authority and Its Subversion," *Glyph*, 8 (1981), 40–61; Richard Helgerson, *The Elizabethan Prodigals* (Berkeley: Univ. of California Press, 1976) and *Self-Crowned Laureates* (Berkeley: Univ. of California Press, 1983); the essays of Louis Adrian Montrose: e.g., " 'Eliza, Queene of shepheardes,' and the Pastoral of Power," *English Literary Renaissance*, 10 (1980), 153–82; and Jonathan Goldberg, *James I and the Politics of Literature* (Baltimore: Johns Hopkins Univ. Press, 1983).

[3] W. Jackson Bate, *The Burden of the Past and the English Poet* (Cambridge, Mass.: Belknap/Harvard Univ. Press, 1970), p. vii.

[4] Harold Bloom, *The Anxiety of Influence* (New York: Oxford

Univ. Press, 1973). In his *Map of Misreading* (New York: Oxford Univ. Press, 1975), Bloom maintains the same focus on post-Enlightenment "strong" poetry, though he does discuss Milton and his precursors.

[5] Said, *Beginnings*, op. cit.

[6] Peter Dronke has raised this issue in his study of poetic innovation and experimentation in Latin works of the eleventh and twelfth centuries, *Poetic Individuality in the Middle Ages* (Oxford: Clarendon, 1970). With him I ask, "Is not *this* the anachronism, reading into the text what the writer has not intended, in order to comply with modern notions of the medieval [and, I would add, the Renaissance]?" (p. 196).

[7] Alice Miskimin, in *The Renaissance Chaucer* (New Haven: Yale Univ. Press, 1975), examines Chaucer's poems as works "concerned with the question of the autonomy of the imagination, which becomes Sidney's, Spenser's and Shakespeare's question as well" (p. 3). John Guillory's *Poetic Authority: Spenser, Milton, and Literary History* (New York: Columbia Univ. Press, 1983) shares my interest in the tension between the autonomous imagination and the authority of "some superior power" in the Renaissance, although his focus on the "anxieties of the *religious* poet" (p. 11) yields broad categories (Scripture vs. literature, divine vs. human) that are not the specific concern of my study. David Quint's *Origin and Originality in Renaissance Literature* (New Haven: Yale Univ. Press, 1983), which analyzes the conflict between claiming participation in a transcendent origin of truth and positing the text as an exclusively human creation, examines the "topos of the source" in several Continental and English writers. Although these last two books appeared too recently for me to incorporate them substantially into my own, I have tried to indicate in footnotes (particularly in my chapter on *The Faerie Queene*) the similarities and differences that characterize our approaches.

Chapter I

[1] As reported by John of Salisbury, *Metalogicon*, ed. C. C. J. Webb (Oxford: Oxford Univ. Press, 1929), Bk. III, Ch. 4 (p. 136).

[2] Gilbert Highet, *The Classical Tradition: Greek and Roman Influences on Western Literature* (New York: Oxford Univ. Press, 1949), p. 261.

[3] J. W. H. Atkins, *English Literary Criticism: The Medieval Phase* (Cambridge: Cambridge Univ. Press, 1943), p. 70; Russell Fraser, *The Dark Ages and the Age of Gold* (Princeton: Princeton Univ. Press, 1973), p. 17; Ernst Robert Curtius, *European Literature and the Latin Middle Ages*, trans. Willard R. Trask (1953; rpt. New York: Harper and Row, 1963), p. 119.

[4] A. J. Minnis, *Medieval Theory of Authorship: Scholastic Literary Attitudes in the Later Middle Ages* (London: Scolar, 1984), p. 12. Minnis's study of the academic prologues to the *auctores* studied in the schools and universities between 1100 and 1400 contributes much to our understanding of the complexity of evolving attitudes of and toward authors in the Middle Ages.

[5] See the series of notes on "Standing on the Shoulders of Giants" in *Isis*—George Sarton, *Isis*, 24 (1935–36), 107–09; R. E. Ockenden, *Isis*, 25 (1936), 451–52; Raymond Klibanksy, *Isis*, 26 (1936–37), 147–49—and Foster Guyer's "The Dwarf on the Giant's Shoulders," *MLN*, 45 (1930), 398–402, for some brief comments on the phrase as conveying a general belief in the "idea of progress." Yet these critics have been cautious about associating even this broad formulation with the twelfth century: Guyer claims that John of Salisbury uses the comparison "in praise of the ancients" (399), and Klibansky calls the identification of the moderns as dwarfs (as opposed to their position on the giant's shoulder) "the really characteristic point of Bernard's dictum" (148). I want also to acknowledge here Robert Merton's *On the Shoulders of Giants: A Shandean Postscript* (New York: Free Press, 1965), a whimsical and digressive book that attempts to trace the origin of the aphorism from Isaac Newton's use of it in the seventeenth century. Merton, a sociologist, both parodies scholarship and pursues it, but he does make a case for the double-edged nature of the analogy. Cf. R. W. Southern, *The Making of the Middle Ages* (New Haven: Yale Univ. Press, 1953), p. 203. Edouard Jeauneau (in " '*Nani gigantum humeris insidentes*': Essai d'interpretation de Bernard de Chartres," *Vivarium*, 5 [1967], 79–99) has attempted to show, by examining the context of the remark in the twelfth century, that to attribute the idea of

progress and praise of the moderns to the original use of the phrase is anachronistic. In my review of that context, I am suggesting that both aspects of the comparison are insisted upon in the twelfth century and that to ignore this is to ignore the complex issues of authority and authorship that the medieval writer faced.

[6] Brian Stock, *Myth and Science in the Twelfth Century: A Study of Bernard Silvester* (Princeton: Princeton Univ. Press, 1972), p. 6. As my analysis below indicates, however, what Stock sees as a "*blend* of tradition and innovation" in Bernard's metaphor (p. 10), I see more as a conflict.

[7] Cf. Merton, pp. 40–41.

[8] *The* Metalogicon *of John of Salisbury: A Twelfth-Century Defense of the Verbal and Logical Arts of the Trivium*, trans. Daniel D. McGarry (Berkeley: Univ. of California Press, 1955), Bk. III, Ch. 4 (p. 167). All further references to this edition will be cited parenthetically within the text, by book, chapter, and page number.

[9] The quoted phrase is from Curtius's *European Literature in the Latin Middle Ages*, p. 53. R. W. Southern has recently attempted to show that the significance of the school of Chartres has been exaggerated; see his *Medieval Humanism and Other Studies* (Oxford: Blackwell, 1970), pp. 61–85, and "The Schools of Paris and the School of Chartres," in *Renaissance and Renewal in the Twelfth Century*, ed. Robert Benson and Giles Constable (Cambridge, Mass.: Harvard Univ. Press, 1982), pp. 113–37.

[10] On the latter see Hans Liebeschutz, *Medieval Humanism in the Life and Writings of John of Salisbury* (London: Warburg Inst., 1950), pp. 11–15, 90–91, and John O. Ward, "The Date of the Commentary on Cicero's 'De Inventione' by Thierry or Chartres (ca. 1095–1160?) and the Cornifician Attack on the Liberal Arts," *Viator*, 3 (1972), 219–73, esp. p. 221, n. 3. These two aspects of the *Metalogicon* go hand in hand, as these writers suggest, because John felt that "the purpose of academic study is to make a man really fit for his activities in Church and State" (Liebeschutz, p. 91).

[11] On this aspect of the "dichotomy" of attitude in the *Metalogicon*, see James J. Murphy, *Rhetoric in the Middle Ages: A*

History of Rhetorical Theory from Saint Augustine to the Renaissance (Berkeley: Univ. of California Press, 1974), pp. 112, 129.

[12] Jeauneau's discussion (see note 5) of John of Salisbury addresses this issue as well and covers some similar material, but we reach different conclusions. Although Jeauneau initially acknowledges that the *Metalogicon* comes out neither clearly for nor clearly against the ancients or moderns (p. 83), he concludes nevertheless that in general, John of Salisbury (among other twelfth-century writers) insisted primarily on the gigantic stature of the ancients rather than on the higher position of the moderns (though without preaching a cult of servility to antiquity) (p. 98).

[13] Although John is speaking specifically of dialectic and logic in this section, not of grammar (or literature), his remarks can be considered as general comments on the relative authority of the ancients and moderns (especially since the *Metalogicon* attempts to reconcile grammar and dialectic).

[14] John was not unwilling, however, to exploit the advantages of association with the *auctores* by engaging in the medieval habit of inventing *auctores* for his work. See Janet Martin's "Uses of Tradition: Gellius, Petronius, and John of Salisbury," *Viator*, 10 (1979), 57–76, for a discussion of John's invention of the *Institutio Traiani*, attributed to Plutarch, as a "pseudo-classical authority and framework for the political ideas he wanted to recommend to his contemporaries" in the *Policraticus*, which left him "free to develop his own ideas" under a "protective cloak of classical authority" (pp. 63, 66).

[15] See Brian Stock's discussion of the "ancients and moderns" in *The Implications of Literacy: Written Language and Models in Interpretation in the Eleventh and Twelfth Centuries* (Princeton: Princeton Univ. Press, 1983), pp. 517–19, where he mentions several other writers who made an appeal for such recognition of the moderns.

[16] See, e.g., Vives's contention (in *De Causis Corruptarum Artium*) that the analogy between dwarfs and giants is a "false and fond similitude": "Neither are we dwarfs, nor they giants, but we are all of one stature, save that we are lifted up somewhat higher by their means" (Foster Watson, Introd., *Vives: On*

Education, a trans. of *De Tradendis Disciplinis* [1913; rpt. Totowa, N.J.: Rowman and Littlefield, 1971], p. cv; quoted in part by Fraser, p. 18); and George Herbert's version of the aphorism in his list of *Outlandish Proverbs*: "A Dwarfe on a Gyants shoulder sees further of the two" (*The Works of George Herbert*, ed. F. E. Hutchinson [Oxford: Clarendon, 1941; rpt. 1972], p. 322).

[17] Cf. Highet, e.g., who claims that Bernard's "objection" "was taken up and turned round, wittily though falsely, by the partisans of the modern side in the Battle of the Books" (p. 267), and Miskimin, *The Renaissance Chaucer*, who implies that the dual interpretation of the dwarf-giant analogy is primarily a Tudor phenomenon, reflecting "the proud Elizabethan sense of advance beyond the predecessors" (p. 18).

[18] Robert Burton, *The Anatomy of Melancholy*, ed. Holbrook Jackson (London: Dent, 1932), Vol. I, p. 25. (Didacus Stella was a sixteenth-century Spanish biblical commentator.) All further references to this work will be cited parenthetically within the text.

[19] Merton notes this immediate context as well (pp. 5–7).

[20] Cf. Stanley Fish on the unreliability of authority in Burton's *Anatomy* in *Self-Consuming Artifacts: The Experience of Seventeenth-Century Literature* (Berkeley: Univ. of California Press, 1972), pp. 303–52 passim, and Joan Webber, *The Eloquent "I": Style and Self in Seventeenth-Century Prose* (Madison: Univ. of Wisconsin Press, 1968), p. 86.

[21] For probing analyses of the relationship between Burton and his persona Democritus Junior, see esp. Webber, pp. 80–96, and Ruth A. Fox, *The Tangled Chain: The Structure of Disorder in the* Anatomy of Melancholy (Berkeley: Univ. of California Press, 1976), pp. 214–35.

[22] This history is documented by Fox, *The Tangled Chain*, pp. 223–26, and I am here summarizing the details of her account.

[23] At the beginning of the Preface, he acknowledges the custom of writers who use "the name of so noble a philosopher as Democritus, to get themselves credit, and by that means the more to be respected," but claims, " 'Tis not so with me" (p. 15).

[24] *Elizabethan Critical Essays*, ed. G. Gregory Smith (London: Oxford Univ. Press, 1904; rpt. 1971), Vol. II, pp. 200–01. I have

modernized the *u/v* spelling. All further references to this work will be cited parenthetically within the text.

[25] *John Milton: Complete Poems and Major Prose*, ed. Merritt Y. Hughes (Indianapolis: Odyssey/Bobbs-Merrill, 1957), p. 720. All further quotations from Milton are from this edition.

[26] I am drawing here upon Anna K. Nardo's excellent analysis of Milton's characteristic pun on *license* and the important distinction he makes between license and liberty (in *Milton's Sonnets and the Ideal Community* [Lincoln, Neb.: Univ. of Nebraska Press, 1979], pp. 70–73).

[27] Horace, *Satires, Epistles, and* Ars Poetica, trans. H. Rushton Fairclough (London: Loeb Classical Library, 1926), pp. 450 (Latin), 451 (English trans.).

[28] Ibid., p. 451.

[29] Quintilian, *Institutio Oratoria*, trans. H. E. Butler (London: Loeb Classical Library, 1921), Vol. I, p. 153.

[30] The phrase is borrowed from Murphy, *Rhetoric in the Middle Ages*, p. 33. My discussion is indebted to Murphy's examination of the *metaplasm*, particularly in relation to Donatus and Isidore of Seville.

[31] Ibid., p. 33.

[32] Ibid., p. 186; Curtius, *European Literature*, p. 44.

[33] Quoted from the translation in *Classical and Medieval Literary Criticism: Translations and Interpretations*, ed. Alex Preminger, O. B. Hardison, Jr., and Kevin Kerrane (New York: Ungar, 1974), p. 378.

[34] Gascoigne, for example, in his *Certayne Notes of Instruction* (1575), advising writers to use accepted English syntax and idiom and to avoid foreign mannerisms, admits that "yet sometimes . . . the contrarie may be borne, but that is rather where rime enforceth, or *per licentiam Poeticam*, than it is otherwise lawfull or commendable." "This poeticall licence," he goes on to explain, "is a shrewde fellow, and covereth many faults in a verse; it maketh wordes longer, shorter, of mo sillables, of fewer, newer, older, truer, falser; and, to conclude, it turkeneth all things at pleasure . . ." (*Elizabethan Critical Essays*, ed. Smith, pp. 53–54; I have modernized the *u/v* spelling).

[35] Gabriel Harvey, *Foure Letters and certaine Sonnets, especially touching Robert Greene* . . . (1592), Bodley Head Quartos, ed.

G. B. Harrison (New York: Dutton, 1923), Second Letter, p. 15. I have modernized the *u/v* spelling.

[36] Ibid., Third Letter, p. 49.

[37] *The Works of Francis Bacon*, ed. James Spedding, Robert Ellis, and Douglas Heath (London: Longman, 1859), Vol. III, p. 343.

[38] Ibid., p. 284.

[39] Hannah Arendt, "What Is Authority?," in *Between Past and Future* (1954; rpt. New York: Viking, 1968), pp. 91–141.

[40] Ibid., pp. 92, 93, 97.

[41] Ibid., p. 97.

[42] Ibid., p. 98.

[43] Ibid., pp. 122–23.

[44] Laurence B. Holland, "Authority, Power, and Form: Some American Texts," *Yale English Studies*, 8 (1978), 2–3. Cf. Said, *Beginnings*, pp. 83 ff., who analyzes the relationship between the words *authority* and *author* and suggests that true, absolute authority is silent, whereas only contingent authority speaks or is written.

[45] Minnis, *Medieval Theory of Authorship*, p. 10.

[46] Ibid., p. 81.

[47] Ibid., pp. 25–26. See also p. 157, where he notes that Vincent of Beauvais, for example, would use the term *actor* to refer to opinions of his own or of the "modern doctors."

[48] *Il Convito: The Banquet of Dante Alighieri*, trans. Elizabeth Price Sayer (London: Routledge, 1887), Bk. 4, Ch. 6 (p. 181).

[49] Ibid., pp. 181–82.

[50] Ibid., p. 182.

[51] Minnis, *Medieval Theory of Authorship*, p. 10. Minnis is referring to the term as it is used in a literary context; Dante's immediate context here is philosophical, and the relation of philosophy to government, though of course the *Convivio* was conceived and structured as a series of commentaries on his own *canzoni*.

Chapter II

[1] Hawes, *Pastime of Pleasure*, ed. William Edward Mead, EETS os 173 (London: Oxford Univ. Press, 1927), ll. 1324–30 (p. 55). I have modernized the *f/s* spelling.

[2] John S. P. Tatlock, *The Development and Chronology of Chaucer's Works*, Chaucer Society, 2nd ser., 37 (1907), p. 38.

[3] See Murray Wright Bundy, " 'Invention' and 'Imagination' in the Renaissance," *Journal of English and Germanic Philology*, 29 (1930), 535–45, for an analysis of how Hawes's use of the term *inventio* in the *Pastime of Pleasure* (especially in ll. 701 ff.) exemplifies a transitional phase during which the medieval notion of rhetorical *inventio* was united with the "imagination" of the psychologist, leading ultimately to the association of invention with originality, "a creation by the poet himself."

[4] Chaucer, *The House of Fame*, ll. 314, 429. All quotations from Chaucer are from *The Works of Geffrey Chaucer*, ed. F. N. Robinson, 2nd ed. (Boston: Houghton Mifflin, 1957).

[5] A. C. Spearing, *Medieval Dream-Poetry* (Cambridge: Cambridge Univ. Press, 1976), pp. 2, 4, mentions this as a characteristic of dream poetry.

[6] Cicero, *De Divinatione* II.lx.124, in *De Senectute, De Amicitia, De Divinatione*, trans. William Armistead Falconer (London: Loeb Classical Library, 1923), p. 511.

[7] Quoted by Werner Wolff, *The Dream: Mirror of Conscience* (New York: Grune & Stratton, 1952), p. 14; see pp. 1–64 for his comprehensive summary of the history of dream theory from antiquity through the twentieth century.

[8] Ibid., p. 20.

[9] Cicero, *De Divinatione* II.lxxi.146; Loeb ed. p. 535.

[10] Ibid., II.lxxi.147; Loeb ed., p. 535.

[11] See Constance B. Hieatt, *The Realism of Dream Visions: The Poetic Exploitation of the Dream-Experience in Chaucer and His Contemporaries* (The Hague: Mouton, 1967), pp. 23–33, for a review of medieval dream theory, especially that of Macrobius and John of Salisbury. Spearing (pp. 6–23) also reviews the tradition of visions that lay behind the medieval dream poem; see esp. his discussion of Macrobius (pp. 8–11). For a discussion of medieval dream lore as treated by medical science (e.g., Galen and Avicenna), astrologers, and theologians, see Walter Clyde Curry, *Chaucer and the Mediaeval Sciences* (New York: Oxford Univ. Press, 1926), pp. 195–218.

[12] Macrobius, *Commentary on the Dream of Scipio*, trans. and ed. William Harris Stahl (New York: Columbia Univ. Press, 1952), Bk. I, Ch. 3 (p. 88).

[13] Ibid., p. 90.

[14] John of Salisbury, *Policraticus*, trans. Joseph B. Pike, *Frivolities of Courtiers and Footprints of Philosophers* (Minneapolis: Univ. of Minnesota Press, 1938), Bk. II, Ch. 15 (p. 79). All further references to this work will be cited parenthetically within the text, by book, chapter, and page number.

[15] See Spearing, p. 25, and Sheila Delany, *Chaucer's* House of Fame: *The Poetics of Skeptical Fideism* (Chicago: Univ. of Chicago Press, 1972), p. 38.

[16] *The Romance of the Rose*, trans. Harry W. Robbins, ed. Charles W. Dunn (New York: Dutton, 1962), 1.1–6, 10–13 (p. 3).

[17] Delany, p. 39. See also H. L. Levy, "As myn auctor seyth," *Medium Ævum*, 12 (1943), 25–39, for a useful discussion of this medieval convention and its gradual disintegration as writers, particularly Chaucer, began to make distinctions between the realm of poetry and the realms of history and science. Cf. also Miskimin, pp. 133–34.

[18] *Romance of the Rose* 85.149–60 (pp. 392–93). Spearing also notices a shift in the status of the dream, but he sees its final classification as more clearly defined: "the transformation of the dream from a vision of truth to a mere fantasy, from a *visio* (in Macrobius's terms) to an *insomnium*" (p. 40). Cf. also Hieatt, pp. 30–31.

[19] See Winthrop Wetherbee, *Platonism and Poetry in the Twelfth Century: The Literary Influence of the School of Chartres* (Princeton: Princeton Univ. Press, 1972), pp. 261–66, and George Economou, *The Goddess Natura in Medieval Literature* (Cambridge, Mass.: Harvard Univ. Press, 1972), pp. 120–21.

[20] See Donald R. Howard, *The Three Temptations: Medieval Man in Search of the World* (Princeton: Princeton Univ. Press, 1966), esp. pp. 175–78.

[21] Morton W. Bloomfield, Piers Plowman *as a Fourteenth-Century Apocalypse* (New Brunswick, N.J.: Rutgers Univ. Press, 1961), p. 3; Elizabeth D. Kirk, *The Dream Thought of* Piers Plowman (New Haven: Yale Univ. Press 1972), p. 200.

[22] *Piers the Plowman*, ed. Walter W. Skeat (1886; rpt. London: Oxford Univ. Press, 1969), B Text, Passus VII, ll. 143–164 (pp. 246–48).

[23] For an elaboration of this point, see Hieatt, p. 20, and Bloomfield, *Piers Plowman*, p. 13.

[24] See, e.g., Bloomfield, *Piers Plowman*, pp. 11–12. Bloomfield asserts that "It is quite clear that the dream was a favorite literary device because it bespoke a revelation, a higher form of truth." Although he is willing, briefly, to admit that "Dream theory in antiquity and the Middle Ages recognized, of course, that there were false or misleading dreams, and much ingenuity was expended in attempting to set up criteria for distinguishing the true from the false," he soon dismisses this fact and maintains, nevertheless, that "In any case, as a literary device, the dream must be understood as a vehicle for the delivery of truth, although not necessarily of divine origin."

[25] Delany, p. 44.

[26] J. A. W. Bennett, *Chaucer's* Book of Fame: *An Exposition of "The House of Fame"* (Oxford: Clarendon, 1968), p. 180.

[27] See, e.g., Spearing, pp. 74–76, esp. p. 76: "one possibility of the dream-framework for Chaucer was that, like a real dream, it could liberate the mind from the demands of causal and rational coherence, so as to open creative opportunities of a different kind"; and Delany, p. 44: "the composition, like the dream, becomes a free-floating quantum of creative or psychic energy . . . The process of dreaming becomes nearly synonymous with the creative act." Cf. also Hieatt, p. 18, and Robert B. Burlin, *Chaucerian Fiction* (Princeton: Princeton Univ. Press, 1977), pp. 26–27.

[28] I do not mean this in the strictly allegorical sense advocated by James Winny (*Chaucer's Dream-Poems* [New York: Harper & Row, 1973]), who interprets the dream vision as an explicit analogue to the workings of the poet's creative imagination. Cf. Spearing's point, that "the dream-poem becomes a device for expressing the poet's consciousness of himself as poet and for making his work reflexive" (p. 6).

[29] Morton W. Bloomfield, "Authenticating Realism and the Realism of Chaucer," in *Essays and Explorations: Studies in Ideas, Language, and Literature* (Cambridge, Mass.: Harvard Univ. Press, 1970), p. 184.

[30] Spearing, p. 75. Although Spearing earlier makes a point similar to mine—that the dream vision "is a poem which does not take for granted its own existence, but is continuously aware of its own existence and of the need, therefore, to justify that existence . . ." (p. 5)—he here concludes that the ambiguous

status of the dream offered the medieval poet a "way out of his dilemma" (p. 74), whereas I am claiming that this reinforced, rather than resolved, the problem.

[31] Chaucer, *The House of Fame*, l.1886.

[32] For a useful analysis of the term *tydynges*, see R. C. Goffin, "Quiting by Tidings in the *Hous of Fame*," *Medium Ævum*, 12 (1943), 40–44, who suggests that it means not only "news" but also "tale," "story," thus reinforcing the idea of the poem as a search for a source of new poetic material. On this, see also Bennett, p. xi, and Robert J. Allen, "A Recurring Motif in Chaucer's *House of Fame*," *Journal of English and Germanic Philology*, 55 (1956), 393–405. The theme of authority in this poem has often been noted as well; my purpose here is to explore in detail the intricate maneuverings of the authorial stance in relation to it.

[33] Commentary on the Chaucerian narrator is extensive and complex; my point here is simply that the narrator, whose character is not consistent even within a single poem, occupies several different positions of control over the meaning of his experience and of his poem. We need not resolve these various stances, or opt for either the naive narrator–ironic author or the ironic narrator–ironic author concept. The issue of the Chaucerian narrator is important to examine as an index to the author's effort to test, explore, and define the relationship between poet and poem, not simply as an index to the author's personality. See Burlin, pp. 27–29, and Thomas Garbaty, "The Degradation of Chaucer's 'Geffrey,' " *PMLA*, 89 (1974), 97–104.

[34] The appeal that God "turne us every drem to goode" is repeated at the beginning and toward the end of this survey of dream lore (ll. 1, 57–58). However, unlike the Browneian "O Altitudo" that is reached after reason has been exhausted, the statement seems to be trivialized by its refrainlike repetition at these particular points of the discussion, carrying the tone of a convenient and offhand gesture rather than a serious affirmation of faith or appeal to God.

[35] The uniqueness of the dream is reiterated in a similar manner in the proem to Bk. II, ll. 509–17).

[36] Delany traces and discusses the use of these two accounts in a very helpful section of her book (pp. 48–57). I should note here my own position in relation to Delany's study of the poem.

Delany's thesis throughout is that in the *House of Fame*, Chaucer confronts "the unreliability of traditional information" (p. 2), which is "multivalent and contradictory" (p. 25), thereby making choice between the different traditions impossible. The difficulty, according to Delany, is transcended by a "fideistic appeal to God or to Christ" (p. 37) that frees the poet from the constraints of these traditions. Although I agree with (and am indebted to) Delany on many points when she discusses the difficulty presented by traditional material, I cannot agree with the direction of resolution toward which she moves. The traditional guides in the poem are problematical not only because they contradict each other but also because they cannot always accommodate the personal demands of the poet. He is not simply testing traditions against each other; he is testing various traditions against his own claims and status as author. It is this that I find to be a "structural pattern" in the *House of Fame*: not a move away from unreliable conventions to the "highest authority" of God (p. 41), but a move away from unreliable conventions to test the extent of the authority of the individual, a move that is constantly reversed and reenacted.

[37] The term is used by Delany (pp. 50, 57), who also notes the similarity to the situation in the Proem; however, she sees the relevance of the legend as residing simply in the "exemplary value" of the dual tradition and not in its "specific content."

[38] George Kennedy, *The Art of Persuasion in Greece* (Princeton: Princeton Univ. Press, 1963), p. 15; my emphasis.

[39] These are, of course, the opening lines of Virgil's *Aeneid*.

[40] Allen points to the constant use of this refrain as a reminder that "our impressions come from an act of artistic creation" (p. 396). Both Allen (pp. 396–97) and Bennett (p. 36) briefly note the narrator's shift from describing wall engravings to presenting his own poetic narrative. Donald K. Fry ("The Ending of the *House of Fame*," in *Chaucer at Albany*, ed. Rossell Hope Robbins [New York: Franklin, 1975], pp. 27–40), also notices what he calls Chaucer's "plot manipulation" but claims, in a very different interpretation from mine, that Chaucer deliberately emphasizes his narrator's "distortion" of a "well-known source for his own purposes" to show us "how authors contribute to the faulty transmission of knowledge from the past" (p. 30).

[41] It is interesting, in this context, that the proverb chosen and so labeled was not a common one. See Albert C. Baugh's note in his edition of *Chaucer's Major Poetry* (New York: Meredith, 1963), p. 32, note to l. 289.

[42] This line of demarcation is pointed out by Robinson in his note, p. 780.

[43] I am not claiming that we are meant to assume the narrator's awareness of this shift in sources; I do think, however, that Chaucer was conscious of how he was employing the two versions of the story and that he structured his narrator's activity relative to his sources so that it works on more than one level. Several layers, hard to separate, operate here: Chaucer's relation to his poem and to his sources in Virgil and Ovid, the narrator's relation to the legend engraved in the Temple, to Virgil and Ovid (whom he cites as sources for his audience to check [ll. 378–79]), and to his dream and narrative. I am not convinced that Chaucer would have intended his audience to unravel all the threads and identify each one; what seems important here is that all contexts coexist and are simultaneously available, creating a rich texture in which the author's relation to authoritative texts (the ones so designated by tradition and the one—his own—he attempts to sanction as such) can appear in its various possible forms and thereby indicate the necessarily tentative, extremely difficult nature of the task of selecting one out and following it through to its end.

[44] William Wilson, "Exegetical Grammar in the *House of Fame*," *English Language Notes*, 1 (1963–64), 244–48, sees the desert as a "state of creative sterility" outside "the magic circle of poetry," which he identifies with the Temple and the story on its walls (pp. 247–48); similarly, Spearing contends that "the landscape conveys a sense of the failure of creative power" (p. 85). My point is that the poet's creative activity has already been stifled by the account of the Dido and Aeneas legend found in the Temple and that the desert presents a new *potential* for creative autonomy.

[45] For some interesting discussions of the significance of the narrator's disinterest in the eagle's scientific knowledge and his confidence in the poetry of the *auctores*, see Allen, p. 400; J. L. Simmons, "The Place of the Poet in Chaucer's *House of Fame*,"

Modern Language Quarterly, 27 (1966), 131; and the counterarguments, concerning whether the distinctions between these two realms are blurred or maintained, of Levy, pp. 33–34, and Wolfgang Clemen, *Chaucer's Early Poetry*, trans. C. A. M. Sym (1963; New York: Barnes and Noble, 1964), p. 98.

[46] See, e.g., Paul G. Ruggiers, "The Unity of Chaucer's *House of Fame*," *Studies in Philology*, 50 (1953), 16-29, rpt. in *Chaucer Criticism*, Vol. II, ed. Richard J. Schoeck and Jerome Taylor (Notre Dame: Univ. of Notre Dame Press, 1961), 261–74: "This is Boethius' own lesson well learned, that inner peace cannot be bought on the world's terms. To it Chaucer adds a faith in oneself, which in his words suggests a faith in God-given talents" (p. 269). See also Stanley E. Fish, *John Skelton's Poetry* (New Haven: Yale Univ. Press, 1965), pp. 55–56.

[47] Cf. Delany, p. 103.

[48] Miskimin, p. 67.

[49] Allen, p. 404; Delany, p. 113.

[50] See, e.g., Bennett, p. 184: "What little we are told of this commanding figure suggests an *auctor* with the prestige of the Africanus of the *Parliament* . . ." See also Ruggiers, pp. 271–72; Simmons, p. 134; Winny, pp. 109–110.

[51] *Troilus and Criseyde* I.65, I.87; cf. Bennett, p. xiii.

[52] Most readers assume that the *House of Fame* is unfinished, and many take it upon themselves to supply the ending that the poem lacks. The response of George Kittredge, *Chaucer and His Poetry* (Cambridge, Mass.: Harvard Univ. Press, 1915), p. 107, is revealing in its honesty: "I am glad the *House of Fame* is unfinished, for this gives me a chance to guess at the story that should conclude it." Guesses have ranged from identifying the "man of gret auctorite" as Boethius (Ruggiers, p. 270) to suggesting that the poem was to serve as an introduction to a group of stories (John M. Manly, "What Is Chaucer's *House of Fame*?" in *Anniversary Papers of G. L. Kittredge* [Boston: Ginn, 1913], p. 77). These attempts to provide conclusions to the *House of Fame* reveal the desire for the resolution that is not present in the poem—one that, however, the poem itself indicates is unavailable. (On this, see Spearing, p. 88, and Delany, p. 109, who see the unfinished quality as appropriate to the unresolvable issues presented. See also Kay Stevenson, "The Endings

of Chaucer's *House of Fame*," *English Studies*, 59 [1978], 10–26, who suggests that the patterns in the poem "make authoritative resolution of tension in the house of Rumor implausible and unnecessary" [p. 15].) In response to David Bevington's remark that "It is unfair to Chaucer's inventive power to suppose him unable to supply . . . an ending" ("The Obtuse Narrator in Chaucer's *House of Fame*," *Speculum*, 36 [1961], 289), I would add that it is also unfair to Chaucer's inventive power to suppose him unable to *refrain* from supplying the ending that would resolve all the issues in the poem. As Bertrand Bronson writes in "Chaucer's *House of Fame*: Another Hypothesis," *Univ. of California Publications in English*, 3 (1934), 190: "There is something, when one comes to consider it, highly suspicious about the poem's breaking off just at the crucial point of the narrative. It has the air not of chance but of deliberate intent." The abrupt cessation of the poetic voice just as the "man of gret auctorite" appears is a highly significant feature of this poem, and it actually supplies a conclusion that fits with the tone of the rest of the poem. Cf. Fry, who argues that the poem breaks off as "a deliberate fragment" (p. 28), though as my discussion below suggests, I find the issue far more complex than Fry, who concludes that Chaucer "mistrusted his *auctoritees*" and "trusted only his own art" (p. 39).

[53] For some concurring opinions about the status of the "man of gret auctorite," see Miskimin, p. 76; Delany, p. 108; Fry, p. 38; Laurence Eldredge, "Chaucer's *Hous of Fame* and the *Via Moderna*," *Neuphilologische Mitteilungen*, 71 (1970), 119; A. Inskip Dickerson, "Chaucer's *House of Fame*: A Skeptical Epistemology of Love," *Texas Studies in Language and Literature*, 18 (1976), 177–78.

[54] My comments on the *Parliament of Fowls* are indebted to H. M. Leicester, Jr., "The Harmony of Chaucer's *Parlement*: A Dissonant Voice," *Chaucer Review*, 9 (1974–75), 15–34; see in particular his point that "it is clear that *Natura's* control of the Valentine's day project is minimal" (p. 27).

[55] This moment is singled out by Leicester, (pp. 29–30), who provides an excellent discussion of the conflict between the individual voice and traditional "truths" in this poem. Leicester's analysis of these traditional truths, which can only be

affirmed "if individuals will agree . . . to self-conscious self-limitation," has influenced my reading of the ending of the *House of Fame.*

[56] See Joseph Anthony Mazzeo, "St. Augustine's Rhetoric of Silence: Truth vs. Eloquence and Things vs. Signs," in his *Renaissance and Seventeenth-Century Studies* (New York: Columbia Univ. Press, 1964), pp. 1–28. On this concept see also Richard McKeon, "Rhetoric in the Middle Ages," in *Critics and Criticism,* ed. R. S. Crane (Chicago: Univ. of Chicago Press, 1952), pp. 260–66, and Marcia L. Colish, *The Mirror of Language: A Study in the Medieval Theory of Knowledge* (New Haven: Yale Univ. Press, 1968), pp. 47–49.

[57] McKeon, p. 266.

Chapter III

[1] Skelton, *The Bowge of Courte,* in *The Poetical Works of John Skelton,* ed. Alexander Dyce (London: Thomas Rodd, 1843), Vol. I. I have modernized all *u/v* spelling.

[2] Fish, *John Skelton's Poetry,* p. 74.

[3] Cf. Spearing, *Medieval Dream-Poetry,* who notes that the narrator finds that "his dream provided him with 'mater of to wryte' " (p. 200).

[4] Boccaccio, *Genealogia Deorum Gentilium* XIV.13, and Harington, *A Preface, or rather a Briefe Apologie of Poetrie,* prefixed to his translation of *Orlando Furioso.*

[5] Sidney, *An Apology for Poetry,* in *Elizabethan Critical Essays,* ed. Smith, Vol. I, p. 184. I have modernized all *u/v* and *i/j* spelling. All further references to this work will be incorporated parenthetically within the text.

[6] William Nelson, *Fact or Fiction: The Dilemma of the Renaissance Storyteller* (Cambridge, Mass.: Harvard Univ. Press, 1973), p. 8. On the relation among fiction, fact, and lying in Sidney's *Apology,* see also A. C. Hamilton, *The Structure of Allegory in* The Faerie Queene (Oxford: Clarendon, 1961), pp. 19–29, and Walter R. Davis, *Idea and Act in Elizabethan Fiction* (Princeton, N.J.: Princeton Univ. Press, 1969), pp. 28–44, both of whom stress the primacy of the fiction as fiction.

[7] Nelson, p. 8.

[8] Nelson, p. 54. See also his discussion of the moral justification of fiction, pp. 49–54.

[9] See Baxter Hathaway's *Marvels and Commonplaces: Renaissance Literary Criticism* (New York: Random House, 1968), esp. pp. 55–56, 88 ff. concerning the "marvelous" nature of invented ideal worlds that stand in contrast to verisimilar imitations of fallen everyday experience full of accidents and contingencies.

[10] Cf. A. Leigh DeNeef's formulation of this crux in the *Apology* (in "Rereading Sidney's *Apology*," *Journal of Medieval and Renaissance Studies*, 10 [1980], 158–59), as well as his suggestive comments on the text's metaphoric nature, which necessarily distorts the Idea (176 ff.).

[11] This point is noted by Hathaway, p. 94.

[12] On the function of poetic fictions and their effect in and on the actual world, see Davis, pp. 40–44, concerning poetry as a mediator between the ideal and the actual; Ronald Levao, "Sidney's Feigned *Apology*," *PMLA*, 94 (1979), 223–33, on the nature of the poet's autonomous "Idea" and Sidney's insistence that it be attached to some ethical and didactic application; and, more generally, Harry Berger, Jr., "The Renaissance Imagination: Second World and Green World," *Centennial Review*, 9 (1965), 36–78, whose provocative discussion of the reflexive relation in the Renaissance between fact and fiction, actuality and imagination, identifies a temporarily self-sufficient fiction that fulfills itself only in a "return to life." I am attempting here to delineate and clarify the problematic consequences this relationship produces for the poet's autonomy and the status of his fiction as they are defined elsewhere in the *Apology*.

[13] See Geoffrey Shepherd's notes to this passage in his edition of the *Apology* (London: Nelson, 1965), p. 217, and O. B. Hardison, Jr., "The Two Voices of Sidney's *Apology for Poetry*," *English Literary Renaissance*, 2 (1972), 94, both of whom liken Sidney's idea of knowledge here to Bacon's empiricism.

[14] George Chapman, Pref., *The Odyssey*, in *English Literary Criticism: The Renaissance*, ed. O. B. Hardison, Jr. (Englewood Cliffs, N.J.: Prentice-Hall, 1963), p. 206.

[15] Boccaccio, *Genealogia Deorum Gentilium* XIV.12, in *Boccaccio on Poetry*, trans. and ed. Charles G. Osgood (Indianapolis: Bobbs-Merrill, 1956), pp. 59–60.

[16] Michael Murrin, *The Veil of Allegory* (Chicago: Univ. of Chicago Press, 1969), who bases his definition of allegory on this idea that it conceals rather than reveals truth, also describes allegory as a form doomed to failure, but the terms he uses are, I think, misleading. He contends that "His [the poet's] failure was the inevitable result of his own intentions. . . . He could not achieve his ends except with a few people" (p. 94), whereas I would say that, given the Boccaccian premise, this "failure" was, for the allegorical poet, success: he *did* achieve his ends when he reached only a few people.

[17] Boccaccio, p. 60.

[18] Murrin acknowledges that this theory of allegory is problematic in that it requires the poet to deal with two different types of audiences (the elite few vs. the many), and therefore "simultaneously to reveal and not to reveal his truth" (p. 168), but he fully accepts the idea that this dual function was achieved by cloaking the truth in the veils of allegory. My point is not simply that the poet had a difficult "double purpose" but that the terms of these two purposes are mutually exclusive and that the "solution" reinforces the contradictions raised by the defense of this mode.

[19] Henry Peacham, *The Garden of Eloquence* (1577), facsimile ed. (Menston, Eng.: Scolar, 1971). All quotations are from this edition. I have modernized the *u/v* and *f/s* spelling.

[20] Cf. Mark Caldwell, "Allegory: The Renaissance Mode," *ELH*, 44 (1977), 580–600, who notes that in general the rhetorical tradition demands that "the link between *litera* and hidden sense [be] explicit" (p. 584).

[21] Puttenham, *The Arte of English Poesie*, facsimile ed. (Kent, Ohio: Kent State Univ. Press, 1970), Bk. 3, Ch. 18 (p. 196). I have modernized the *u/v* and *f/s* spelling. All further references to this work will be incorporated parenthetically within the text, by book, chapter, and page.

[22] Cf. Rosemond Tuve and Angus Fletcher who, in different ways, posit a virtually endless allegorical form that can never fully articulate the ideals it aspires to. See Tuve's *Allegorical Imagery* (Princeton: Princeton Univ. Press, 1966), esp. her remarks on "entrelacement" and how allegory provides "the opportunity to realize, re-realize, and realize again, the full import

of something we can only lamely point to by its abstract name," a "something" she also refers to as "unparaphrasable meaning" (p. 370), and Fletcher's *Allegory: The Theory of a Symbolic Mode* (Ithaca: Cornell Univ. Press, 1964), p. 177: "all analogies are incomplete, and incompletable, and allegory simply records this analogical relation in a dramatic or narrative form."

[23] Isabel G. MacCaffrey, *Spenser's Allegory: The Anatomy of Imagination* (Princeton: Princeton Univ. Press, 1976), p. 56. See her helpful discussion of Puttenham's analysis of allegorical language and of the necessity of employing words that "have been warped away from accuracy of meaning" (pp. 54–57). Cf. also Maureen Quilligan, *The Language of Allegory: Defining the Genre* (Ithaca: Cornell Univ. Press, 1979). Although to my mind she too facilely dismisses as "erroneous" the traditional definitions of allegory that were passed down to the Renaissance (p. 29), Quilligan, arguing for the polysemous nature of language as the defining feature of allegory, writes that "The allegorical poet simply asks in narrative form what the allegorical critic discursively affirms; are my words lies, or do they in fact thrust at the truth?" (pp. 46–47).

[24] The question of the intended accessibility of allegory to its audience is a vexed issue among modern critics. Murrin claims that allegory, as articulated by men like Boccaccio and practiced by Spenser, is "preeminently an obscure form of poetry" (p. 8) whose special mode of rhetoric is defined by the "concealment of truth" (p. 116). He opposes this allegorical mode to the oratorical tradition of Puttenham, Sidney, and Jonson, which demands perspicuity in order to reach its audience. Yet Edward A. Bloom, "The Allegorical Principle," *ELH*, 18 (1951), 163–190, contends generally that "Obscurity . . . has never been considered an attribute of allegory" (p. 174). He claims that the most common feeling through the eighteenth century was that "the moral equivalent of the allegorical fabrication must be transparent enough to be accessible" (p. 175) and that those who conceived of the genre as cloudy and guilty of deliberate concealment *objected* to it on those grounds. He places Sidney and Spenser together in the first group as advocates of transparent allegory and opposes them to Puttenham as a critic of allegory as a means of deception. MacCaffrey, on the other hand, claims

that Puttenham *defends* allegorical "darkness" as essential to the ultimate revelation of truth (pp. 39–40, 45). And Peter Berek, "Interpretation, Allegory, and Allegoresis," *College English*, 40 (1978), 117–32, claims that there were two kinds of allegory, one based on a principle of clarity and one based on concealment; he groups Puttenham, Sidney, Jonson, and Spenser together in the former category, Boccaccio in the latter. As these disparate commentaries indicate, there is some confusion as to whether Renaissance allegory was generally intended to keep remote or expose its truths. The attempts of these different critics to place various Renaissance writers in opposing camps yield contradictory groupings, implying that perhaps we cannot make such a division. I am suggesting here that the tension between these two theories was inherent *within* various definitions of poetry and allegory in the Renaissance and that it is a simplification to propose that only one notion prevailed either for the period in general or for any one writer in particular. The movement between identifying allegory as a mode that retreats from and one that reaches towards the world of its readers seems endemic to almost all theories of allegory and, I think, defines a crucial characteristic of the mode and a crucial issue for the poet.

[25] Cf. John M. Steadman, *The Lamb and the Elephant: Ideal Imitation and the Context of Renaissance Allegory* (San Marino, Calif.: Huntington Library, 1974), who makes a different point when he locates a tension (and ultimately an historical movement) between allegory—"the principle of *hiding* an ideal meaning in a symbolic and metaphorical vehicle that is overtly incredible, improbable, and unreasonable"—and mimesis—"the principle of *revealing* or illustrating the ideal meaning through an exemplary narrative or figure that is consistently credible, probable, rational, and 'like the reality' " (p. 89). I am speaking not of a conflict between allegory and an Aristotelian verisimilitude that could picture ideals through a mode that was "like the reality" but rather of an internal conflict within allegorical theory between the ideals and a mode that participated "*in* the reality."

[26] Levao, p. 227.

[27] Cf. Murrin's remarks on "audience control" (pp. 171–74),

and MacCaffrey (pp. 33 ff.) on how allegory imitates man's epistemological dilemma and mode of discursive reasoning, though I cannot then share her conviction that allegory makes special reference to the "accessibility of truth" (p. 40) and achieves "intensity and completeness of illumination" (p. 41).

[28] See MacCaffrey, p. 54: "All the important characteristics of allegory reflect and encourage an awareness that we live in a radically deformed universe. In this fallen world, the universe of discourse is as severely warped as any other aspect of reality." Cf. Murrin, p. 87.

[29] I must distinguish my position from some other studies that have posited the idea that allegory is a mode that approximates rather than opposes actuality. MacCaffrey's provocative book, for example, which I have cited throughout, seems to be making a point similar to mine: allegory "belongs to the fallen world," it "imitates everyday, ill-informed experience," and it is "a mirror of our fallen condition" (pp. 34, 41, 45); MacCaffrey consequently explains its "dark conceit" as essential to its reader's understanding. However, she presents these characteristics as the accepted *premises* of allegory, and she ignores that the approximation of the reader's world and his epistemological mode evolves from a radically different motive. The original impulse is to define allegory as a mode that not only is detached from but also deflects man's fallen experience, and this original but ultimately defeated impulse must be recognized to understand the nature of allegorical writing, its relation to that experience, and the poet's relation to his text.

A similar problem exists in Edwin Honig's extremely suggestive *Dark Conceit: The Meaning of Allegory* (Evanston, Ill.: Northwestern Univ. Press, 1959), which not only explores the thesis that "The literary allegory does not oppose a realistic account of the universe" (p. 179) but also examines the question of authority in allegory, positing that "the allegorical writer seeks . . . a new or recoverable authority for the active imagination" (p. 94). Yet, while Honig claims that "The poet's role . . . will remain completely subservient to the historian's unless it can be shown that he creates something quite different from a version of history" (p. 109), he also contends that for allegorists to establish authority in fiction "the challenge was to map out the relation of the unknown country of allegory to the known countries and conditions of contemporary actuality" (p. 103), without commenting on the potential conflict between two such statements. For example, when speaking of *The Faerie Queene*, although he initially emphasizes Spenser's insistence in his letter to Raleigh on the distinction between the methods of the poet and the historian (p. 95), Honig generalizes on "Spenser's way of facing the question of authority—which meant giving substance to his work by having his figures, each with various roles, participate in both his fairyland and in the real world" (pp. 103–04). As the next section of this chapter will discuss, this was not Spenser's solution; rather, it was the problem he never resolves. Studies like these view allegory as a mode that incorporates the real and the ideal, whereas I claim that for the Renaissance allegorist, this proximity of allegory and actuality would likely signal the demise or loss of the ideal.

[30] *The Faerie Queene* VII.vi.1.1. All quotations are from the Variorum Edition, ed. Edwin Greenlaw, Charles Osgood, and Frederick Padelford, 10 vols. (Baltimore: Johns Hopkins Press, 1932–57), and will be incorporated parenthetically within the text.

[31] In the following discussion I make some points similar to A. Leigh DeNeef's in *Spenser and the Motives of Metaphor* (Durham, N.C.: Duke Univ. Press, 1982). We diverge on a basic issue, however, and our arguments seem finally to move in opposite directions. DeNeef claims in general that Spenser ultimately tries to defend his "metaphoric faerieland" against the "literal-minded" perception and portrayal of absolute disparities between, for example, ideality and reality, fiction and fact; while acknowledging that Spenser occasionally succumbs to such disparities to ward off abuses, he focuses primarily on Spenser's attempt to "void their dichotomies" (p. 106). I have focused instead on the alternative perspective: the ways in which such dichotomies—however impossible to maintain—are crucial to his conception of his art, and on how their dissolution or "accommodation" is seen as a threat or challenge to rather than a justification of it.

[32] See Carol Kaske, "Spenser's Pluralistic Universe: The View from the Mount of Contemplation" in *Contemporary Thoughts on*

Edmund Spenser, ed. Richard C. Frushell and B. J. Vondersmith (Carbondale: Southern Illinois Univ. Press, 1975), pp. 127–30, although her ultimate point is that, while the worlds are contradictory and irreconcilable, "Spenser refuses to take sides as to which is the truer picture of man"; and Harry Berger, Jr., *The Allegorical Temper: Vision and Reality in Book II of Spenser's* Faerie Queene (New Haven: Yale Univ. Press, 1957), p. 104.

[33] Denis Donoghue, *Thieves of Fire* (New York: Oxford Univ. Press, 1974), p. 17. For other discussions of the Promethean origins of the Faery race, cf. Thomas P. Roche, Jr., *The Kindly Flame: A Study of the Third and Fourth Books of Spenser's* Faerie Queene (Princeton: Princeton Univ. Press, 1964), pp. 34 ff.; Berger, pp. 107–11; and Kaske, pp. 127–29. Michael O'Connell, *Mirror and Veil: The Historical Dimension of Spenser's* Faerie Queene (Chapel Hill: Univ. of North Carolina Press, 1977), who also sees the Promethean story's significance in its emphasis on the fact that the ideal of Faeryland is created and fictional, does not see the negative implications that imbue this story of creation.

[34] See Roche, who maintains that "Faeryland is the ideal world of the highest, most virtuous *human* achievements" (p. 45), and O'Connell, who contends that since the ideal "risks becoming remote," Faeryland "must be placed in the context of a world in which earthly, infected will is still very active" (p. 81). Cf. Stephen Greenblatt's comments—though not specifically tied to the chronicles—on the explicit "createdness" of Spenser's art as it reveals itself to be an aesthetic "ideal image" of ideology and denies "possession or embodiment of reality," in *Renaissance Self-Fashioning* (Chicago: Univ. of Chicago Press, 1980), pp. 189–92.

[35] Inconclusiveness, incompleteness, and loose ends are the hallmark of *The Faerie Queene*. The narrator's self-consciousness and uneasiness about this unfinished quality, and his Ariostan transitions that call attention to it, are newly developed, beginning in Book III, firmly established by Book IV. See Allan H. Gilbert, "Spenser's Imitations from Ariosto," *PMLA*, 34 (1919), 225–32.

[36] The only narrative problems the narrator announces in this section are ones of compression; that is, he invokes the con-

ventional difficulty of the self-proclaimed transcriber, without incurring any of the risks of the maker. In my characterization of the narrator's role here (or, rather, my acceptance of his characterization of it) I am taking a perspective different from that of critics who see the account of the river marriage as a triumph of the poet and of poetry, an explicit image of the power of art and of the imagination at work. (See, e.g., A. Bartlett Giamatti, *Play of Double Senses: Spenser's* Faerie Queene [Englewood Cliffs, N.J.: Prentice-Hall, 1975], pp. 130–33.) Cf. MacCaffrey, who calls the triumph of concord in the river marriage "profoundly natural" (p. 118).

[37] Paul Alpers also notes this point in *The Poetry of* The Faerie Queene (Princeton: Princeton Univ. Press, 1967), p. 121.

[38] By "Faeryland" I am referring here to the fictional world of the poem itself in which its major characters move and act and in which their stories unfold.

[39] O'Connell also discusses the Book II chronicles together with the river marriage, but he sees in Book IV a withdrawal into "a more private poetic world" and claims that in this section, "Spenser is broadening his claim for the importance of his own poetic world" (p. 89). Once we take into account the details of the Florimell and Marinell story and their relation to the Thames and the Medway, however, this sort of position becomes, I think, difficult to maintain. DeNeef, on the other hand, does make the comparison between the two stories, noting that the river marriage "excludes man from its celebration . . . The world of man, it seems, . . . [is] controlled by destructive and deforming forces." However, since DeNeef follows critics like Giamatti in seeing the river marriage as the *poet's* victory, he concludes with an interpretation opposite to mine: "Freely ranging in his own golden world, the poet seems to have turned away from the brazen one . . ." (p. 124). In my reading the final cantos of Book IV attest to the poet's *inability* to range freely within his own golden world and to the identification (rather than the detachment) of the poet's created world and the brazen world. The general shape of my discussion of the Book IV "marriages" is in several respects closer to Jonathan Goldberg's (in *Endlesse Worke: Spenser and the Structures of Discourse* [Baltimore: Johns Hopkins Univ. Press, 1981]), who ana-

lyzes the ways in which the "Authority of an Other" shapes the text as the "world and word meet" variously in these episodes (pp. 122–65). In his view, however, if I understand him correctly, this "other" that shapes the text is to be associated with the reality of political, courtly, and social *fictions* that manifest royal power and authority, whereas I argue that it is the reality of actual existence—limited, disorderly, disruptive—that shapes the poet's art, which is differentiated from the orderly union represented by the natural river marriage.

[40] Harry Berger's comments on Spenser's "dynamic" conception of the *disconcordia concors* and the lack of absolute resolutions or triumphs in *The Faerie Queen* (in "The Spenserian Dynamics," *Studies in English Literature,* 8 [1968], 1–18) inform my discussion in this chapter, though again I cannot accept the rigid distinctions he erects between human and faery.

[41] Roger Sale, *Reading Spenser: An Introduction to* The Faerie Queene (New York: Random House, 1968), pp. 171–81. (Sale locates this change beginning in Book V.)

[42] Harry Berger, Jr., "The Prospect of Imagination: Spenser and the Limits of Poetry," *Studies in English Literature,* 1 (1961), 98, 97, 104. See also Richard Neuse, "Book VI as Conclusion to *The Faerie Queene,*" *ELH,* 35 (1968), 329–53.

[43] Other critics have recently suggested that Spenser is confronting the implications of his own poetic mode more than the exigencies of the actual world, but they focus on different characterizations of this mode. See, e.g., Madelon Gohlke, "Embattled Allegory: Book II of *The Faerie Queene,*" *English Literary Renaissance,* 8 (1978), 123–40, who argues that Spenser is "coming to terms with the limitations of his genre," a genre that is "based on" a "*separation*" between moral ideals and fallen human reality. While I agree that this separation exists on one level, my point is that Spenser confronts the limitations imposed by a genre that ultimately enables him neither to *maintain* that separation in his delineation of the relation between Faeryland and the actual world nor, therefore, to create his own image of those ideals. Cf. also Deneef, who claims that in Book V, Spenser concedes that his bleak vision is caused not by the condition of reality but by his own poetic strategy, which has emphasized a *disparity* between reality and fiction and which

he thus must refashion to accommodate the fiction, the ideal, to the actual (pp. 124–33). More similar to my position is Goldberg's claim that the increased sense of problems in narration in Book IV is caused not by external pressures but by "the nature of narration itself," which denies closure (p. 29), and his focus on how "the social situation of the text is figured in the text" (p. 128). However, Goldberg's commitment to the Barthesian notion of a "writerly text" as an explanatory principle leads him in a different direction: for example, when he discusses the dissolution of boundaries between what is "inside" and "outside" the text, stating that the poem cannot be said to replicate " 'uninvented reality' " if reality is itself a "textual invention" (p. 29). When he does address the issue of the "external as the nontextual" (p. 123) later in his book, in the chapter entitled "The Authority of the Other," his primary concern there is, as I mentioned in note 39, with the social and courtly *fictions* of royal power that surround Queen Elizabeth, and he still speaks of the "*fiction* that the text has a referent" outside of itself (p. 126). Though I do not share Goldberg's specifically political definition of the Authority and hence of the producer of the text, or his view of what constitutes the "reality" that is external to and brought into the text, I do very much share his perception that (in the terms of his discourse) "When the word enters society, or society enters the word, the word is destroyed" (p. 169).

[44] Berger, "Prospect of Imagination," p. 97. Berger seems to waver between saying that the poet withdraws from the actual into Faerie and that the actual invades Faerie. In his later version of this article, "A Secret Discipline: *The Faerie Queene*, Book VI," in *Form and Convention in the Poetry of Edmund Spenser*, ed. William Nelson, English Institute Essays (New York: Columbia Univ. Press, 1961), pp. 35–75, he emphasizes the conflict between these two ideas and concludes that "the vision must be bounded and shaped by the sense that it is not reality; and it must yield to reality at last" (p. 75).

[45] For some further comments on the dramatic mode of narration in the *Mutability Cantos*, see Paul Alpers, "Narration in *The Faerie Queene*," *ELH*, 44 (1977), esp. 36–37.

[46] Cf. Guillory, *Poetic Authority*, who also sees the *Mutability*

Cantos as an "analysis of authority." Basically, however, although we cover some similar ground—for example, in our views of Jove as a figure of authority and of the narrator's difficulty in representing Nature—we have taken fundamentally different perspectives on the issue of authority. Guillory sees Spenser seeking first a "sacred" authority and, when this is displaced, deferring instead for sanction to a "secular" authority defined as a "rhetoric of textual continuity" (p. 66) "established by the citation of literary *auctors*" (p. 64). I am more concerned throughout with how Spenser confronts the question of the authority of his *individual* authorial voice as he weighs it against external authorities that fail to provide adequate, practicable, or acceptable alternatives. Cf. also Quint's discussion, in *Origin and Originality*, of how Spenser "makes allegory dependent upon and thus vulnerable to the events of history" (p. 31). Quint's focus, however, on *The Faerie Queene* II–VI as an "Elizabeth-centered historical allegory" leads him to see in the *Mutability Cantos* a "*loss* of a historical source of allegorical meaning" (a source that he finds emblematized in the Thames and Medway marriage) and hence "an end to Spenser's historical allegory" (pp. 149–66).

[47] This idea is implied in Harry Berger's discussion of the "desymbolizing" of the gods in "The *Mutabilitie Cantos*: Archaism and Evolution in Retrospect," in *Spenser: A Collection of Critical Essays* (Englewood Cliffs, N.J.: Prentice-Hall, 1968), pp. 146–76, and is further explored in Lewis Owen's discussion of the "humanization" of the gods, particularly of Jove, in "Mutable in Eternity: Spenser's Despair and the Multiple Forms of Mutabilitie," *Journal of Medieval and Renaissance Studies*, 2 (1972), 49–68. Along related lines see James Nohrnberg's comments on the "dialectical" relationship that exists between the characters in the *Mutability Cantos* in *The Analogy of* The Faerie Queene (Princeton: Princeton Univ. Press, 1976), esp. pp. 741–48.

[48] It is worth noting here that the literary relationship between the Goddess Nature and man has not been an easy one, for their history together, though initially benign and harmonious, is characterized by tensions, complaints, rejections, and misunderstanding. The concept of the goddess likely derived from Plato's doctrine in the *Timaeus*, where he posits a world-soul

as intermediary and harmonizing principle between the ideal and material worlds. It was in that middle ground that the twelfth-century School of Chartres positioned its Goddess Natura—as a figure of reconciliation and as guardian of human stability. But for the Chartrian writers, the relationship between Nature and man was tenuous and easily disturbed; their poetry generally presents Nature as an ideal and authority but also reveals both the difficulty of fulfilling that ideal on the human level and of relying upon Nature's authority. In *De Mundi Universitate,* Bernardus Silvestris suggests the possibility of man's divergence from the principle of harmony and her laws (in physical and psychological terms); in *De Planctu Naturae,* Alain de Lille enlarges this to include the inadequacies of human language, expression, and comprehension: men are unable to read Nature properly, and Nature cannot always successfully explain her principles to the human mind. Jean de Meun (*Roman de la Rose*) decreased the distance between man and Nature, not by depicting the ideal and authoritative Nature as available to man, but rather by humanizing the goddess and thereby divesting her of her status. In the *Parliament of Fowls,* Chaucer sets up Nature as an authoritative guide to whom men (and humanlike birds) can appeal, but in the end, she, like all the other proposed authorities, is found inadequate. Man and Nature rarely acheived the easy harmonious relationship that the goddess herself was intended to preserve, and it was this long tradition of a problematic, disintegrating association between the two that Spenser inherited. See the studies of the Nature tradition by Economou (*The Goddess Natura in Medieval Literature*) and Wetherbee (*Platonism and Poetry,* as well as his Introduction to his translation of *De Mundi Universitate* [New York: Columbia Univ. Press, 1973]) to which my summary here is indebted.

[49] On Nature as reconciling oppositions, see William Nelson, *The Poetry of Edmund Spenser* (New York: Columbia Univ. Press, 1963), pp. 306–07, and Kathleen Williams, *Spenser's World of Glass: A Reading of* The Faerie Queene (Berkeley: Univ. of California Press, 1966), pp. 230–31. See also Milton Miller, "Nature in *The Faerie Queene,*" *ELH,* 18 (1951), 191–200, for a more general analysis of the way in which contradictions between

dual orders of things are only apparent ones, made by men who do not see "from the point of view of the whole" (196).

[50] See note 48. For other analyses of the implications of Spenser's citation of *auctors*, cf. Miskimin, pp. 41–43, and Guillory's counterargument, pp. 61–64.

[51] See Milton Miller's formulation of this concept, which he traces throughout *The Faerie Queene*, and C. S. Lewis, *The Allegory of Love* (London: Oxford Univ. Press, 1936), p. 356.

[52] Cf. Owen's probing analysis of how Mutability challenges the distinction between terrestrial and celestial regions and Nature fails to reaffirm it.

[53] Other critics who note that the narrator does not accurately summarize what Nature has said include Owen, 59–60; Thomas Greene, *The Descent from Heaven* (New Haven: Yale Univ. Press, 1963), p. 322; Ricardo Quinones, *The Renaissance Discovery of Time* (Cambridge, Mass.: Harvard Univ. Press, 1972), p. 288; and Judith Anderson, *The Growth of a Personal Voice* (New Haven: Yale Univ. Press, 1976), p. 201. See also Michael Holahan, "*Iamque opus exegi*: Ovid's Changes and Spenser's Brief Epic of Mutability," *English Literary Renaissance*, 6 (1976), 244–70, who discusses how Nature's "judgment of reconciliation" is portrayed as "beyond the view of poet and poem," providing evidence of the poet's "imaginative humility"; and MacCaffrey, pp. 430–31, on Spenser's acknowledgment of the "limits of his art."

[54] See W. B. C. Watkins, *Shakespeare and Spenser* (Princeton: Princeton Univ. Press, 1950), p. 72.

[55] On this point see Joanne Field Holland, "The Cantos of Mutabilitie and the Form of *The Faerie Queene*," *ELH*, 35 (1968), rpt. in *Critical Essays on Spenser from* ELH (Baltimore: Johns Hopkins Press, 1970), pp. 256–57; Berger, "Archaism and Evolution," pp. 147–48; and Nelson, pp. 294 et passim. My list of nonresolutions follows Nelson's. Particularly relevant to my discussion is Berger's comment on how, in the *Cantos*, Nature's reconciliation triggers a new "moment of anxiety" in the final stanzas.

[56] See Holland, p. 257; for later formulations see, e.g., DeNeef, p. 93, and Guillory, p. 28.

[57] I here differ from those studies that also discuss the implications of the *Cantos* for the poet's own writing but con-

clude that they signal the end of Spenser's allegory. Cf., e.g., MacCaffrey, who sees the poem ending in silence and passivity, claiming that "The poet ceases to be the active shaper of an allegorical poem" (p. 432), and Nohrnberg's analysis of the "end of the technique of Spenser's allegory" (pp. 776–77). My point is that in those last stanzas, the poet resurrects his role as an allegorical poet and resurrects his allegorical poem, while at the same time expressing his uneasiness with and desire to be released from them. From a different but complementary perspective, see Patricia Parker's discussion (in *Inescapable Romance* [Princeton: Princeton Univ. Press, 1979], pp. 54–64) of the potentially endless structure of the *romance* form of *The Faerie Queene* and its relation to the "dilation" of which Nature speaks at the end of the *Mutability Cantos.*

Chapter IV

[1] Petrarch, *Familiare* 23.19, trans. Morris Bishop, *Letters from Petrarch* (Bloomington: Indiana Univ. Press, 1966), pp. 198–99.

[2] Ibid., p. 199.

[3] Thomas Greene, *The Light in Troy: Imitation and Discovery in Renaissance Poetry* (New Haven: Yale Univ. Press, 1982), p. 96 (and also in his earlier article "Petrarch and the Humanist Hermeneutic," in *Italian Literature: Roots and Branches*, ed. Giosi Rimanelli and Kenneth Atchity [New Haven: Yale Univ. Press, 1976], p. 212). Greene's book is the central and definitive study of Renaissance imitation. Of the four types of imitation he delineates, I am most concerned here with the one he terms "dialectical" (closely allied with the "heuristic"), which reflects a "resistance or ambivalence toward imitation" and which reveals, in Greene's words, that "The process called imitation was . . . a field of ambivalence, drawing together manifold, tangled, sometimes antithetical attitudes, hopes, pieties, and reluctances within a concrete locus" (pp. 43–45). I should add that Greene's book was published after this chapter was written (this section on imitation was presented at the 1981 MLA convention). The basis for my selection of texts throughout my discussion of "creative imitation" rests predominantly on my attempt to explore the difficulties of that concept as they are

exposed by the use of a certain image as a metaphor for it. Although imitation of other authors is my primary concern, notions of mimesis and imitation of ideals, ideas or universals are included for examination mainly when a writer associates them himself—and, in that context, to show how these types of imitation raised similar issues concerning autonomy and authorship, and how they are often brought into a discussion to balance or support some of the claims that "creative imitation" made for the poet's freedom. For some other recent discussions of imitation in the Renaissance, see Jerome Mazzaro, *Transformations in the Renaissance English Lyric* (Ithaca: Cornell Univ. Press, 1970), pp. 73–107; Howard C. Cole, *A Quest of Inquirie: Some Contexts of Tudor Literature* (Indianapolis: Pegasus/Bobbs-Merrill, 1973), pp. 216–26; Marion Trousdale, "Recurrence and Renaissance: Rhetorical Imitation in Ascham and Sturm," *English Literary Renaissance*, 6 (1976), 156–79; Margaret Ferguson, "The Exile's Defense: Du Bellay's *La Deffence et illustration de la langue francoyse*," *PMLA*, 93 (1978), 275–89; William Kerrigan, "The Articulation of the Ego in the English Renaissance," in *The Literary Freud: Mechanisms of Defense and the Poetic Will*, ed. Joseph H. Smith (New Haven: Yale Univ. Press, 1980), pp. 261–308; and G. W. Pigman III, "Versions of Imitation in the Renaissance," *Renaissance Quarterly*, 33 (1980), 1–32, discussed in note 9 below. The majority of modern critics (Pigman, Ferguson, and Greene are exceptions) follow the lead of Harold O. White, *Plagiarism and Imitation during the English Renaissance* (Cambridge, Mass.: Harvard Univ. Press, 1935), who claims that the theory of imitation—from classical times through the Renaissance—always safeguarded, provided the means for, and even required originality. In this chapter I try to clarify particular attitudes and problems that explain why this relationship was not easily accepted or achieved.

[4] Petrarch, *Familiare* 23.19; trans. Bishop, pp. 199–200.

[5] Ibid., p. 199.

[6] Greene, p. 98. But cf. his earlier comments on how "the sweetness of otherness constitutes a risk" (pp. 96–97) and his readings of Petrarch's imitative poems, which he finds less "confident" and harmonious and more ambivalent than his letters on imitation.

[7] Petrarch, *Familiare* 22.2; trans. Bishop, p. 183.

[8] Ibid., pp. 182–83.

[9] Cf. Pigman, "Versions of Imitation," who also discusses the digestive metaphor (among others) of imitation. Pigman's analysis of Petrarch is similar to mine in that he too views Petrarch as writing not simply about how to imitate properly but also about the difficulty of doing so. But Pgiman locates the difficulty primarily in the association of "transformative imitation" with "dissimulative imitation"—that is, "when transformation is conceived as the means of hiding a text's relation to its model" (p. 12)—whereas I see the problem inherent in the idea of transformative imitation itself. Furthermore, despite his recognition that reproduction might be a consequence of complete assimilation, Pigman's attempt to catagorize metaphors according to types of imitation leads him to conclude that digestion represents "successful transformations of a model" (p. 3) and that (with the single exception of Cortesi) the metaphor "is always used to support transformative imitation" (pp. 7–8), whereas my analysis reveals that the digestive metaphor both supports and undermines the idea of "transformative"—or creative—imitation. In general though, Pigman's penetrating analysis, particularly of the ambivalence inherent in *eristic* metaphors of emulation, both complements and confirms my discussion.

[10] Erasmus, *Ciceronianus*, trans. Izora Scott in *Controversies over the Imitation of Cicero* (New York: Teachers Coll., Columbia Univ., 1910), Pt. II, p. 123. Also cited by Neil Rudenstine in his discussion of Sidney's anti-Ciceronianism in *Sidney's Poetic Development* (Cambridge, Mass.: Harvard Univ. Press, 1967), esp. pp. 134–40. For some other commentary on Erasmus's theory of imitation, see Greene, pp. 181–84; Pigman, "Versions of Imitation," 8–9, who sees his use of the digestive topos in this passage as "representative"; and Pigman, "Imitation and the Renaissance Sense of the Past: The Reception of Erasmus' *Ciceronianus*," *Journal of Medieval and Renaissance Studies*, 9 (1979), 155–77).

[11] Bacon, "Of Studies," in *The Essays of Francis Bacon*, ed. Samuel H. Reynolds (Oxford: Clarendon, 1809), p. 342. Also quoted by Greene in his article "Petrarch and the Humanist

Hermeneutic," p. 223, n. 23. Bacon's lack of detail may be the result of the fact that his discussion is directed to the general reader—not only to readers who are also poets.

[12] Ben Jonson, *Discoveries*, ed. C. H. Herford, Percy and Evelyn Simpson, in *Ben Jonson* (Oxford: Clarendon, 1947), Vol. VIII, pp. 638–39. For other discussions of Jonson's concept of imitation, see Wesley Trimpi, *Ben Jonson's Poems: A Study of the Plain Style* (Stanford: Stanford Univ. Press, 1963), pp. 41–59; Ira Clark, "Ben Jonson's Imitation," *Criticism*, 20 (1978), 107–27; Greene, esp. pp. 274–78; and Richard Peterson, *Imitation and Praise in the Poems of Ben Jonson* (New Haven: Yale Univ. Press, 1981), esp. 1–43.

[13] Cf. Peterson who, arguing for the "coherence" of Jonson's theory of imitation, explains away this complicating remark as "atypical of Jonson" (p. 6, n. 6). See Greene, who is more sensitive to the ambivalence and tension that informs this passage in particular and the *Discoveries* in general.

[14] Pico, "A Pamphlet on Imitation by Gianfrancesco Pico Addressed to Pietro Bembo," trans. Izora Scott in *Controversies over the Imitation of Cicero*, Pt. II, p. 1. Pico does, of course, proceed to state quite definitely his position.

[15] Castiglione, *The Book of the Courtier*, trans. Charles Singleton (New York: Anchor, 1959), Bk. I, pp. 63–64.

[16] Jonson, *Discoveries*, Vol. VIII, p. 616. Although Jonson is speaking specifically about beginning writers here (juxtaposing them to "grown" writers who "worke with their owne strength"), the conflict cannot be dismissed. Cf. Greene, p. 275.

[17] *Discoveries*, Vol. VIII, p. 586.

[18] Ferguson, p. 276; Kerrigan, pp. 276–77.

[19] Dorothy Connell, *Sir Philip Sidney: The Maker's Mind* (Oxford: Clarendon, 1977), pp. 4–5; Nancy Struever, *The Language of History in the Renaissance* (Princeton: Princeton Univ. Press, 1970), p. 87.

[20] See Struever, who later modifies her position, noting that the reciprocity could also be a "difficult ambiguity" (p. 153). See also Greene, passim, on how, in his fine phrase, the humanist text "had to expose the vulnerability of the subtext while exposing itself to the subtext's potential aggression" (p. 45).

[21] Puttenham, *The Arte of English Poesie*, Bk. I, Ch. 1, pp. 19–20.

[22] MacCaffrey, *Spenser's Allegory*, p. 22; Rosalie Colie, *Paradoxia Epidemica* (Princeton: Princeton Univ. Press, 1966), p. 61.

[23] The substitution of "every thing is set before him" for the "foreine copie or example" does not relieve the burden of reconciliation. Like the narrator in the *House of Fame* who rejects the authority of the eagle's empirical science and logic, momentarily asserting his own authorial voice only to replace it immediately with the authority of the *auctores*, Puttenham's shift to the mimetic concept of imitation merely reinstitutes a different but no less troublesome model. For a discussion of the implications of this shift, particularly as it introduces the problematic relationship between God as Creator of the world and poet as creator of his text, see pp. 136–39.

[24] Hardison, "The Two Voices of Sidney's *Apology*," p. 97.

[25] In the first half of the *Apology*, Sidney is speaking of the poet's freedom from bondage both to Nature and to other authors (the latter especially in his comparison between the historian—"authorising himselfe . . . upon other histories"—and the poet—who has "all . . . under the authoritie of his penne"). I have included his statements about mimesis for their implications about the general nature of a poet's autonomy. The notions of freedom are subverted in *both* contexts in the latter part of the *Apology*.

[26] *Apology*, p. 195: "But these, neyther artificiall rules nor imitative patternes, we much cumber our selves withall."

[27] Hardison, p. 94.

[28] Ibid., 99. Nor can I fully agree with the conclusion of Martin Raitiere, in "The Unity of Sidney's *Apology for Poetry*," *Studies in English Literature*, 21 (1981), 37–57. Rebutting Hardison, Raitiere also argues that no single passage of the *Apology* can be selected as "truly representative of Sidney's critical views" (p. 55). But Raitiere's argument that "the *Apology* possesses a unity that transcends these contradictions" (p. 39)—based to a large extent on a syntactical analysis of crucial passages—maintains what he calls the "paradox" of these voices by making it the "unifying strategy" of the work. Although I agree that the

Apology defines rather than answers the dilemma of Renaissance writers, I find no transcendent unifying structure that "controls" the dilemma in the *Apology*. Along similar lines, see Margaret Ferguson, "Sidney's *A Defence of Poetry*: A Retrial," *Boundary 2*, 7 (1979), 61–95, who claims that the " 'contradiction' between freedom and fettering is a dialectical one woven into the text from beginning to end" (p. 90, n. 31). My point is closer to Levao's contention that the *Apology* acts out "the tensions characteristic of the best Renaissance thought" ("Sidney's Feigned *Apology*," p. 232).

[29] Petrarch, *Familiare* 1.8, trans. Aldo S. Bernardo, *Rerum Familiarium Libri, I–VIII* (Albany: State Univ. of New York Press, 1975), p. 41.

[30] Ibid., p. 41.

[31] Ibid., pp. 41–42.

[32] Petrarch, *Familiare* 23.19; trans. Bishop, p. 200.

[33] Petrarch, *Familiare* 1.8; trans. Bernardo, p. 42.

[34] Rosalie Colie, "Some Paradoxes in the Language of Things," in *Reason and the Imagination: Studies in the History of Ideas 1600–1800*, ed. J. A. Mazzeo (New York: Columbia Univ. Press, 1962), pp. 102–03.

[35] MacCaffrey, p. 22.

[36] See Connell, pp. 1–3, on how Sidney "defines the source and the limits of human creativity in the higher creativity of God" (p. 2). See also Struever, pp. 45–46, 51, 93, for comments on the human/divine artist analogy (in Nicholas of Cusa and Salutati) as an indication that the poet's similarity to the divine Creator lies precisely in his *creative* activities, with unlimited possibilities.

[37] Marvell, "Upon Appleton House," from *The Poems and Letters of Andrew Marvell*, Vol. I, ed. H. M. Margoliouth, 2nd ed. (Oxford: Clarendon, 1952), ll. 441–48.

[38] William C. Johnson, "Spenser's *Amoretti* and the Art of the Liturgy," *Studies in English Literature*, 14 (1974), 48.

[39] J. W. Lever, *The Elizabethan Love Sonnet* (1956; rpt. London: Methuen, 1966), p. 137.

[40] Maurice Evans, *English Poetry in the Sixteenth Century* (London: Hutchinson, 1955), p. 97.

[41] Patrick Cruttwell, *The English Sonnet*, Writers and their Work, No. 191 (London: Longmans, Green, 1966), p. 14.

[42] Nelson, *The Poetry of Edmund Spenser*, p. 85.

[43] Theodore Spencer, "The Poetry of Sir Philip Sidney," *ELH*, 12 (1945), 267.

[44] Donne, "The Triple Foole," from *The Poems of John Donne*, ed. Herbert J. C. Grierson (Oxford: Clarendon Press, 1912).

[45] Raleigh, "Sir Walter Raleigh to the Queen," from *The Poems of Sir Walter Raleigh*, ed. Agnes M. C. Latham (Cambridge, Mass.: Harvard Univ. Press, 1962).

[46] E. C. Pettet, "Sidney and the Cult of Romantic Love," *English*, 6 (1946–47), 235; Robert Kellogg, "Thought's Astonishment and the Dark Conceits of Spenser's *Amoretti*," *Renaissance Papers* (1965); rpt. in *The Prince of Poets: Essays on Edmund Spenser*, ed. John R. Elliott, Jr., (New York: New York Univ. Press, 1968), p. 140.

[47] Rudenstine, *Sidney's Poetic Development*, pp. 46–47. On the relation between poetry and love, see also Connell, *The Maker's Mind*, pp. 9–33.

[48] Petrarch, *Familiare* 22.2; trans. Bishop, p. 183.

[49] Sidney, *The Counteses of Pembroke's Arcadia (The Old Arcadia)*, ed. Jean Robertson (Oxford: Clarendon, 1973), Bk. I, p. 20.

[50] Ibid., Bk. I, pp. 22–23.

[51] *The Poetical Works of William Drummond of Hawthornden*, ed., L. E. Kastner (Manchester: Manchester Univ. Press, 1913), Sonnet 2.

[52] Giles Fletcher, *Licia*, in *Elizabethan Sonnets*, Vol. II. ed. Sidney Lee (1904; rpt. New York: Cooper Square, 1964), Sonnet 1. Unless noted otherwise, quotations from other sonnet sequences (Drayton's *Idea*, Lodge's *Phillis*, and Daniel's *Delia*) are also from this edition.

[53] *William Shakespeare: The Complete Works*, ed. Alfred Harbage, Pelican Text, rev. ed. (London: Penguin, 1969), Sonnet 78. In these sonnets a young man fills the conventional role of the beloved.

[54] The dedicatory sonnet to Sidney's sister implies that she is patron, inspiration, and subject of his sonnet sequence.

[55] A. C. Hamilton, "Sidney's *Astrophel and Stella* as a Sonnet

Sequence," *ELH*, 36 (1969), 73. Cf. his later formulation of this in his *Sir Philip Sidney: A Study of his Life and Works* (Cambridge: Cambridge Univ. Press, 1977), p. 91: "What binds the lover, however, frees the poet: the lady, isolated in her virtue, is his security that his desire to write sonnets will not be frustrated." Cf. also the comments of Peter Cummings, "Spenser's *Amoretti* as an Allegory of Love," *Texas Studies in Language and Literature*, 12 (1970), esp. 169–70, on how the lady's will to remain apart motivates the range of the man's imaginative responses.

[56] Louis L. Martz, *The Poetry of Meditation* (New Haven: Yale Univ. Press, 1954), pp. 269–70.

[57] All quotations from *Astrophil and Stella* are from *The Poems of Sir Philip Sidney*, ed. William A. Ringler, Jr. (Oxford: Clarendon, 1962).

[58] Fish, *Self-Consuming Artifacts*, pp. 198–99.

[59] Colie, *Paradoxia Epidemica*, pp. 195–96.

[60] Rosemond Tuve, in *Elizabethan and Metaphysical Imagery* (Chicago: Univ. of Chicago Press, 1947), suggests that here Sidney is talking about " 'inventing' or finding matter" (p. 39).

[61] The distinction that Fish makes between the authorial positions in "Jordan I" and "Jordan II" seems applicable, up to a point, to the different authorial stances of Herbert and Sidney.

[62] Colie, *Paradoxia Epidemica*, pp. 92–93.

[63] *Essays by Rosemond Tuve: Spenser, Herbert, Milton*, ed. Thomas P. Roche, Jr. (Princeton: Princeton Univ. Press, 1970), pp. 175–76.

[64] Richard Lanham, in "*Astrophil and Stella*: Pure and Impure Persuasion," *English Literary Renaissance*, 2 (1972), 100–15, nicely emphasizes the rhetorical and "persuasive" purpose of the sonnets, but he misses the point when, at the end of his essay, he complains that the sequence lacks a consistent concept of language.

[65] Richard B. Young, "English Petrarke: A Study of Sidney's *Astrophel and Stella*," in *Three Studies in the Renaissance: Sidney, Jonson, Milton*, Yale Studies in English, Vol. 138 (New Haven: Yale Univ. Press, 1958), p. 9.

[66] In the same year that my discussion of the sonnets was originally published as an article (1979), two other studies appeared that examine the relationship between lady and poet in

Astrophil and Stella in complementary terms. See Murray Krieger's analysis of the significance of Stella as the "object of presentation" whose presence in turn creates the poem, in "Poetic Presence and Illusion I: Renaissance Theory and the Duplicity of Metaphor," *Critical Inquiry*, 5 (1979), rpt. in *Poetic Presence and Illusion: Essays in Critical History and Theory* (Baltimore: Johns Hopkins Univ. Press, 1979), pp. 3–27 (see my response to this in " 'What may words say': The Limits of Language in *Astrophil and Stella*," in *Sir Philip Sidney and the Interpretation of Renaissance Culture*, ed. Gary Waller and Michael Moore [London: Croom Helm, and Totowa, N.J.: Barnes and Noble, 1984], pp. 95–109); and Richard McCoy's discussion of autonomy and submission in the "sexual politics" of the sequence in *Sir Philip Sidney: Rebellion in Arcadia* (New Brunswick, N.J.: Rutgers Univ. Press, 1979), pp. 69–109.

[67] The sonnets of the *Amoretti* that I will examine have closer parallels to other sonnets in *Astrophil and Stella* than to those I have discussed; my point is that Spenser never takes the extreme self-effacing posture that Sidney *sometimes* decides to occupy.

[68] Hallett Smith, *Elizabethan Poetry: A Study in Conventions, Meaning, and Expression* (Cambridge, Mass.: Harvard Univ. Press, 1952), p. 166.

[69] Quoted and discussed by Watkins in *Shakespeare and Spenser*, pp. 259–62. Watkins generally agrees with this description of Spenser's poetry (although he notes some tension in the neoplatonic elements of the *Amoretti*), but claims, in opposition to O'Connor, that it is not a basis for disparagement of the sonnets.

[70] Lever, *The Elizabethan Love Sonnet*, pp. 102, 135.

[71] Waldo F. McNeir, "An Apology for Spenser's *Amoretti*," *Die Neueren Sprachen*, 14 (1965), rpt. in *Essential Articles for the Study of Edmund Spenser*, ed. A. C. Hamilton (Hamden, Conn.: Archon, 1972), pp. 524-33, unlike most other critics, claims that Spenser's sonnets are "dramatic in method" (p. 525), but by this he simply means that they employ a strategy of "direct address": "they are spoken by one person to another" (p. 531).

[72] The companion sonnet to this (#71) similarly leaves the identity of the spider ambiguous; it is unclear whether the

spider/artist/captor is the poet or the lady, and indeed, whether the spider is captor or captured.

[73] The nature of Spenser's equivocating poem is highlighted when we compare the context of his statement that the lady "with one word my whole years worke doth rend" to Herbert's "The Collar," in which God's voice enters and, with one word ("Child!"), completely undoes the poet's argument and resistance.

[74] This connection is noted in the Variorum (p. 429); Ronsard's "Astree 11" is cited as a source.

[75] Louis Martz, "The *Amoretti*: 'Most Goodly Temperature,' " in *Form and Convention in the Poetry of Edmund Spenser*, p. 165.

[76] Lanham, "Pure and Impure Persuasion," esp. p. 102.

[77] Here I differ from Young, who claims that Astrophil "deals with the difference between writing poems and stealing kisses, the one an artificial, the other an essential manner of loving" (p. 7). Astrophil identifies the two rather than differentiating them. I am in general agreement with James Finn Cotter's essay, "The 'Baiser' Group in Sidney's *Astrophil and Stella*," *Texas Studies in Language and Literature*, 12 (1970–71), 381–403, which analyzes how Sidney writes within the "baiser" tradition in order to examine, test, critique, and refashion it.

[78] Castiglione, *The Book of the Courtier*, Bk. IV, as quoted by Lever, pp. 122–23. See also Nicolas Perella, *The Kiss, Sacred and Profane* (Berkeley: Univ. of California Press, 1969), pp. 175–80, for a discussion of Castiglione's comments.

[79] For example, Lever, pp. 128–29; Richard Neuse, "The Triumph over Hasty Accidents: A Note on the Symbolic Mode of the *Epithalamion*," *Modern Language Review*, 61 (1966), esp. 164.

[80] Lever, p. 129, makes a similar connection, as does A. Leigh DeNeef in " 'Who now does follow the foule Blatant Beast': Spenser's Self-Effacing Fictions," *Renaissance Papers* (1978), 11–21.

[81] A. Kent Hieatt's numerological analysis (*Short Time's Endless Monument: The Symbolism of the Numbers in Edmund Spenser's "Epithalamion"* [New York: Columbia Univ. Press, 1960]), leads him to conclude that the *Epithalamion* is a symbol not only of one

day, but also of all the days of the year, and also of all time, the "cyclical harmony of an according whole" (p. 54). This distorts, I think, both the tone and the content of the poem. Spenser emphasizes that his "one day" is, in fact, only a single day; the references to the passing of time stress not the representative value of this day as a symbol for all time, but rather the unenduring quality of this day that will, in time, pass away and be replaced by other, less happy days. The envoy does not perform a "recompensing function" (p. 57); the poem is not what Hieatt calls "Short Time's Endless Monument" or "an eternal monument *to* time" (p. 48) but rather a monument "*for short* time." I have similar objections to the optimistic interpretation of A. R. Cirillo, "Spenser's *Epithalamion*: The Harmonious Universe of Love," *Studies in English Literature*, 8 (1968), 19–34, who likewise discovers a "complete harmony celebrated by the poem" and sees "all of nature—the entire universe—as a symphony to wedded love" (p. 20). The world as Spenser depicts it in the *Epithalamion* is antagonistic to his single day of wedded bliss, and the poem tries to protect this "one day" from the external world and the rest of time. See Thomas Greene, "Spenser and the Epithalamic Convention," *Comparative Literature*, 9 (1957), 215–28, who notes that "The poem is unconventional in the repeated expression it gives to the . . . elements which might potentially destroy the joy of the wedding and even the marriage" (p. 226).

[82] Wolfgang Clemen, "The Uniqueness of Spenser's *Epithalamion*," in *The Poetic Tradition: Essays on Greek, Latin, and English Poetry*, ed. D. C. Allen and Henry T. Rowell (Baltimore: Johns Hopkins Press, 1968), praises the poem for its "fusion of heterogeneous material, of disparate notions and motifs" (p. 92); C. S. Lewis also notes this quality of "inclusiveness": the "triumphant fusion of many different elements" (*English Literature in the Sixteenth Century* [Oxford: Clarendon, 1954], p. 373). I see this "inclusiveness" as fundamentally exclusionary, as a diversionary and delaying strategy that provides extra material for the poem as it attempts to forestall its inevitable end.

[83] Greene notes the atypicality of the length of Spenser's *Epithalamion*: "Spenser's stanza is unusual . . . in its number of

recurrences. There are twenty-three stanzas in the *Epithalamion*, whereas the typical Italian *canzone* does not exceed eight or ten" (p. 226). I interpret this unusual length as part of Spenser's attempt to make the poem/day as long as possible.

[84] Chapman, *Hero and Leander*, Sestiad 5, l. 496, from *The Poems of George Chapman*, ed. Phyllis Brooks Bartlett (New York: Modern Language Assn., 1941).

Index